San Diego Christian College
2100 Greenfield Drive
El Cajon, CA 92019

ABC OF PHYSICS
A Very Brief Guide

ABC OF PHYSICS
A Very Brief Guide

Lev Okun

*Institute of Theoretical and
Experimental Physics (ITEP), Russia*

World Scientific

NEW JERSEY • LONDON • SINGAPORE • BEIJING • SHANGHAI • HONG KONG • TAIPEI • CHENNAI

Published by

World Scientific Publishing Co. Pte. Ltd.

5 Toh Tuck Link, Singapore 596224

USA office: 27 Warren Street, Suite 401-402, Hackensack, NJ 07601

UK office: 57 Shelton Street, Covent Garden, London WC2H 9HE

British Library Cataloguing-in-Publication Data
A catalogue record for this book is available from the British Library.

ABC OF PHYSICS
A Very Brief Guide

ISBN-13 978-981-4397-27-8
ISBN-10 981-4397-27-X

Printed in Singapore by World Scientific Printers.

Preface

0.1 For whom this book is written*

Physics is the science describing nature (*phisis* in Greek ($\varphi\upsilon\sigma\iota\varsigma$) means nature), and this book concentrates on its foundations (its "ABC"). The foundation of the physics of the 21st century is the so-called Standard Model of elementary particles of which everything that exists is built in accordance with the laws of the theory of relativity and quantum theory.

It is regrettable that the existing introductions to these sciences are beyond the grasp not only of students but also of the majority of physics teachers, all because of the mathematical complexity of the available textbooks. This little book is a result of my attempt to present briefly and with maximum simplicity the fundamentals of the relativity theory (with its main constant c) and quantum mechanics (with its main constant $\hbar$). I do understand, of course, that I cannot write a popular book which would be comprehensible to readers knowing no physics. Consequently, this book is intended first and foremost for professional physicists (especially for young professors) in the hope that some of them would use the ideas presented in the book for writing simple courses of modern physics and science-popularising books and articles.

In contrast to popular science books, this one makes extensive use of mathematical formulas. However, in contrast to the formulas in university-level textbooks, formulas selected for this book are so simple that the knowledge of elementary mathematics taught at high schools is sufficient for understanding them. I am convinced therefore that many of the pages[1] of the book will be understood even by very young people and should help those who are attracted to physics to gradually expand their scientific horizons. To achieve this, the book ends with a detailed index. There is no doubt that the small size of this brief guide excludes using it as a replacement for a proper textbook.

Elementary particles of matter moving at speeds close to speed of light c and at very small values of angular momentum approaching Planck's quantum $\hbar$, have very unusual properties. However, it is precisely their unusual, counterintuitive nature that makes the world around us what it is.

One of the best scientific descriptions of the world, based on the theory of relativity and quantum mechanics is found in the fundamental monographs written by Stephen Weinberg [1],[2],[3]. However, to present the most essential

---

[1] The corresponding sections and chapters are marked by an asterisk.

elements of these sciences in this small and simplified book, I not only had to select the simplest of formulas but also to refuse mentioning many complex concepts that make the subjects of these and many other treatises of modern physics.

0.2 On the contents of some of the chapters

The fundamental concept of relativity theory is the relation between energy E, momentum $\mathbf{p}$ and mass m of a free particle. this relation, explained in detail in chapter 4, makes it possible to understand within the same framework the properties both of massless particles, such as particle of light — photon, and of particles with nonzero mass, such as electron. According to quantum mechanics, the masses of all particles of one type, e.g. the masses of all electrons in the universe, are strictly identical.

Alas, the equality $E = mc^2$, ubiquitous in popular science literature and interpreted by some as the dependence of the particle mass m on particle velocity v, promotes a false idea of what mass actually is. This false notion gets entrenched even further when people refer to the genuine mass m as "rest mass" and denote it by m_0, thus breaking the link between the theory of relativity and Newtonian mechanics, which is the limiting case of the theory of relativity for low velocities.

The spin (i.e. the intrinsic (proper) angular momentum of a particle) plays the fundamental role in the workings of Nature. (I will explain in chapter 5 that the magnitude of spin is usually expressed in units of $\hbar$.) Particles whose spin is an integer multiple of $\hbar$ are known as bosons. Particles whose spin is a half-integer multiple of $\hbar$ are known as fermions. The behavior of particles in a system consisting of two or greater number of fermions is radically different from the behavior of particles in a system consisting of two or greater number of bosons. For instance, the properties of atoms in the periodic table of elements are a consequence of the electron spin being equal to $\frac{1}{2}\hbar$.

The photon's spin equals $1\hbar$. Electrons interact with each other and with other electrically charged particles by exchanging photons. The theory of interaction of photons with electrons is called Quantum Electrodynamics (QED). QED is the most advanced physics theory, and has been confirmed experimentally to within ten significant digits. Likewise, the exchange of gravitons — massless particles with spin $2\hbar$ — explains gravitation.

In contrast to the approach favored by most textbooks on quantum mechanics, the basis of the presentation in this book is not the relatively complex concept of the wave function of a physical system (such as the atom) but a much simpler concept of an elementary quantum state of the system (atom's energy level) which is characterized by the values of several quantum numbers. See chapter 5.

You will see in chapter 6 that this makes it possible to introduce the amplitude of transition between two quantum states. The amplitude modulus squared gives the probability of transition (decay of an excited state of the atom resulting in the emission of a photon) as a function of time for an ensemble of identical states. Note that the decay of a particular atom is a completely random process, and the moment of time when this decay will occur cannot be predicted in principle. This approach eliminates many of the difficulties of interpretation of quantum mechanics involved in the problem of measurement (reduction of the wave function). On the other hand, this approach makes it possible to highlight those quantum aspects which are not related to the procedure of measurement, and thereby go beyond the confines of defining quantum mechanics as a science which only establishes relations between the results of measurements.

Another specific feature of this book is that the conventional gravitational interaction is treated as a quantum process of exchanging a massless particle of spin 2 known as graviton. Newton's constant G_N together with the constants $\hbar$ and c define the so-called Planck mass $m_P = (\hbar c / G_N)^{\frac{1}{2}} \approx 2.2 \cdot 10^{-5}$ gram. The theory that would describe the gravitational interactions at energies on the order of $m_P c^2$ has not been worked out so far although the correctness of descriptions of gravitation as exchange of gravitons has been confirmed by experiments at all attainable energies. See chapter 16.

0.3 Two beacons*

I wrote this book in the hope that it will prove useful to young people and will help them find their place in this rapidly changing and in many ways incomprehensible world. The main idea I wish to communicate is that the truth does exist and that one can get much closer to it than many people think. The physics holds several simple and inviolable truths that any thinking person can understand. Mankind was able to uncover these truths when studying nature, mostly during the last hundred years.

At first glance, the science has grown, ramified and specialized so much that a non-expert simply cannot hope to understand what its foundations are. The rapid growth of the Internet exposes us to an unprecedented flow of information and misinformation. It is impossible to figure out what is what unless one understands the physical meaning of the constants c and $\hbar$. These two universal constants will help you, like two guiding stars, to find your way to understanding nature. Choosing these most fundamental constants of nature to serve as as your guiding stars in science will allow you to gradually find like-minded individuals and to formulate criteria of truth in specific situations.

Newton compared himself to a boy playing with pebbles on the shore of the great ocean of truths waiting to be discovered. Physics has expanded enormously since then and the area of the known has become vast. Some contemporary scientists are convinced that the realm of the unknown shrank to become smaller than that of the known. I do not know if this is true but I consider it extremely important to separate what has been established with certainty from what only looks plausible.

It seems to me very important to separate the well-established and understood phenomena and properties (such as for example the properties of atoms and their nuclei or the main properties of the solar system) from much more complex issues (such as the birth of the Universe or the behavior of the gravitational interaction at the Planck scale where it is stronger than all other interactions). Such theories as classical mechanics and electrodynamics, or quantum mechanics and field theory will be valid forever. Only certain improvements are possible that would not change the established mathematical relations between physical quantities but would clarify the range of applicability of these relations and justify the choice of the variables they have introduced. In contrast to this, farfetched theoretical extrapolations may not be confirmed by subsequent measurements.

0.4 Acknowledgments

I am grateful to E. G. Gulyaeva and B. L. Okun for their invaluable support during the writing of this book and to Professor K. K. Phua for the invitation to publish this book in English which I received before I began working on the Russian original. I am also grateful to M. Ya. Amus'ya, A. E. Bondar, Yu. B. Danoyan, D. S. Denisov, A. D. Dolgov, V. E. Fortov, J. M. Frere, S. I. Godunov, Yu. M. Kagan, M. O. Karliner, I. B. Khriplovich, A. B. Kojevnikov, O. V. Lychkovsii, N. A. Nekrasov, M. E. Peskin,

N. G. Polukhina, M. B. Voloshin, I. S. Zuckerman and I. I. Zuckerman who read the manuscript and made a number of very useful comments. Their remarks helped me reformulate some statements that were not accurate enough.

This work was supported by a grant of the President of the Russian Federation No. NSh-4172.2010.2, RFBR grant No. 10-02-01398 and contract No. 02.740.11.5158 of the Ministry of Education and Science of the Russian Federation.

Lev Okun
Moscow
February 2012

Contents

Chapter 1

The Fundamentals

1.1   On intuition

The purpose of this book is to explain within a hundred-odd pages how modern elementary particle physics allows us to understand the workings of the surrounding world. Two universal constants lie at the basis of the cosmos: the maximum speed of uniform rectilinear motion c and minimum quantum of rotational and oscillatory motion $\hbar$. The words "rectilinear" and "uniform" seem to point to unbounded, infinite motion through space and to linearity of the time (the philosophy of the Occident). Rotation points to cyclical nature of time and to bounded, finite nature of motion (the philosophy of the Orient).

The matter I wish to present here is not philosophy but the way in which physicists quantitatively describe their observations and experiments using intuition as their tool. The main problem of modern physics is that the concepts of maximum speed and minimum quantum appear extremely anti-intuitive to those who do not work in fundamental physics.

1.2   Space and time*

Our space is three-dimensional: any observer can imagine a system of coordinates — three mutually orthogonal axes: back-to-forward (x), left-to-right (y), bottom-to-top (z). The observer typically places himself at the origin of coordinates.

Imagine an observer following the motion of a particle. He characterizes the position of the particle at the moment of time t by the radius vector $\mathbf{r}$ with its three coordinates: x, y, z. The coordinate system plus time form what we call the frame of reference. All points in space and time are equivalent: both space and time are uniform. In addition, space is isotropic: the three axes can be oriented in an arbitrary manner. Note that under such re-orientations the magnitude of the radius vector remains unchanged: $r \equiv |\mathbf{r}| = \text{const}$.

1.3   Matter and substance*

Any change in the motion or behavior of a particle is a manifestation of elementary (i.e. the smallest and indivisible) particles of matter. We begin with three elementary particles of matter: the photon γ, the electron e and the proton p. We will first see how the simplest hydrogen atom emerges as a result of their interaction. We will then look at other atoms so as to understand how the matter around us emerges as a result of their interaction and formation of increasingly complex systems of interlinked particles.

Then we will go deeper step by step and learn about all other elementary particles.

There is still no universally accepted definition of what should be called 'substance' and what should be called 'matter'. Many authors use the term "material particle" only if the particle has nonzero mass. Suffice it to recall the term "material point" which is used in the literature by many authors (see e.g. Weinberg's book [1]) where they discussed "massive points", that is bodies whose size can be considered negligible in a specific problem. For these authors photons are particles of radiation, not of matter. The debate on what should be called which name will be solved some day. At the moment I will refer to all particles including photons as particles of matter. And I will call atoms and everything built of atoms as substance.

1.4 Motion*

Motion is the displacement of a particle in space. If the motion is translational (i.e. uniform and rectilinear) then the velocity $\mathbf{v}$ of displacement is given by the formula $\mathbf{v} = \mathbf{r}/t$ where $\mathbf{r}$ is the path covered and t is the time the displacement took. The definition of velocity in the case of arbitrary motion will be given a little later.

Chapter 2

Units*

2.1 Standards

Every measurement is a comparison of what is being measured with a known standard. The generally accepted standards with which results of measurements are compared are called units.

2.2 Circle and angles

Angles are measured in degrees (minutes and seconds) or in radians. One degree of arc is the angle subtended at the center of a circle by the arc obtained by dividing the circle into 360 equal parts. The minute of arc: $1'=1°/60$. The second of arc: $1''=1'/60$. We know that the ratio of the circumference of the circle to its radius equals 2π, where $\pi \approx 3.14$. One radian $1 \text{ rad} = 360°/2\pi \approx 57°$ is the angle subtended by the arc whose length equals that of the radius.

2.3 Units of time and length

Units of time are determined by the rotation of the Earth. A day and a night are 24 hours long. One hour = 60 minutes. One minute = 60 seconds. Although one hour and one degree of arc consist of 60 minutes and 3600 seconds, this is the only common feature between the units of angle and time.

Lengths are measured by comparing them against the established length standard. The meter is the typical unit for measuring lengths.

Chapter 3

A Minimum of Mathematics

Mathematics is often said to be the queen and a servant of all the natural sciences. But mathematics is also a powerful obstacle when people are introduced to these sciences. This is why authors of popular science books and articles avoid using mathematical notation. In my opinion, modern physics cannot be explained without mathematics. I will try to limit myself to a minimum of mathematical knowledge. But this minimum is necessary and unavoidable.

3.1 Four operations of school mathematics and the imaginary unit*

Two types of arithmetic operations will be needed here to reach understanding of the laws of nature: 1) addition/subtraction, 2) multiplication/division, and two types of algebraic operations: 3) raising to power and 4) finding the logarithm. It won't be soon that we will need this last operation. We will also need the concept of the imaginary unit $i = \sqrt{-1}$ and of complex numbers.

3.2 Powers of ten*

When studying nature, we constantly come across very large or very small numbers which are conveniently presented as powers of ten. Thus the speed of light is $3 \cdot 10^5$ km/s $= 3 \cdot 10^8$ m/s, and the size of an atom is on the order of 10^{-10} m. Unlike authors of popular science books, I will not use expressions like "one thousandth of one millionth of"; I will write 10^{-9}.

3.3 Prefixes of the powers of ten

Many physical terms include the following prefixes:

Deca	10^1	da	Deci	10^{-1}	d
Hecto	10^2	h	Centi	10^{-2}	c
Kilo	10^3	k	Milli	$10{-}3$	m
Mega	10^6	M	Micro	10^{-6}	μ
Giga	10^9	G	Nano	10^{-9}	n
Tera	10^{12}	T	Pico	10^{-12}	p
Peta	10^{15}	P	Femto	10^{-15}	f
Exa	10^{18}	E	Atto	10^{-18}	a
Zetta	10^{21}	Z	Zepto	10^{-21}	z
Yotta	10^{24}	Y	Yocto	10^{-24}	y

3.4 Differentiation and integration

The velocity of arbitrary motion is determined as the derivative of distance with respect to time: $\mathbf{v} = \mathrm{d}\mathbf{r}/\mathrm{d}t$. The quantity $\mathrm{d}t$ stands for a sufficiently small ("infinitely small") difference between two very close points in time. The quantity d**r** has a similar meaning — the difference between two points in space very close to one another. Here the infinitely small distance between two points in space is divided by the infinitely short interval of time, which gives as a result the finite velocity of motion. This operation is called differentiation. Integration is an operation inverse to differentiation: $\mathbf{r} = \int \mathbf{v} dt$. We won't need these operations just yet. It would be very useful to readers — both familiar with the concepts of differentiation and integration and to complete novices — to browse time and again the remarkable book written by Yakov Borisovich Zeldovich [4].

3.5 Matrices

To learn about quantum mechanics, it will be necessary for us to be introduced to matrices.

Chapter 4

Translational Motion*

4.1 Free particle

The notion of free body or free particle is undoubtedly an idealization — an abstraction, but an exceptionally fruitful abstraction. A particle is described as free if it is not subjected to external forces. If it was at rest then it will continue to be at rest; if it was moving at some velocity, it will keep moving at the same velocity in the same direction. Ordinary bodies around us do stop after a while but this happens because they are not free: other bodies act on them as brakes.

4.2 Maximum velocity c

A remarkable result achieved during the last two-three centuries was the discovery of the fact that nature imposes a maximum on the velocity of translational motion in vacuum, equal to the speed of light $c \approx 300,000$ km/s (to be precise, $c = 299\,792\,458$ m/s; in fact this equality creates a link between the definitions of meter and second).

4.3   Energy and momentum of a particle

The uniform linear motion — i.e. the motion of a free particle — is characterized by two quantities that are conserved in time: its energy E and its momentum $\mathbf{p}$. Energy is a scalar quantity and momentum is a vector quantity characterized by a direction in space. Energy conservation is a consequence of time uniformity while conservation of momentum is a consequence of space uniformity.

Energy is generally defined as the universal measure of all processes and the momentum as the universal measure of all motions. However, these definitions are not specific enough for one to work with them. The evolution of mathematics and physics during the last century was such that the definitions of basic concepts were becoming progressively more difficult while the operations over them and the proofs were getting progressively easier. Energy and momentum are excellent examples of this. It appears that the optimal way of working with them lies in writing the basic formulas for them and gradually finding out how they agree with the concepts familiar to the reader. We shall start with Newton's mechanics.

4.4   Kinetic and potential energy in Newtonian mechanics

The kinetic energy of a free body in Newtonian mechanics is defined as the quantity E_K:

$$E_K = m\mathbf{v}^2/2 = \mathbf{p}^2/2m. \qquad (4.1)$$

If the body is not free but exists in some sort of force field then it possesses a potential energy U in addition to the kinetic energy E_K. For example, a mug on the table has potential energy greater than its potential energy on the floor by a quantity $U = mgh$ where m is the mug's mass, $g \approx 10\,\mathrm{m/s^2}$ is the gravity acceleration on the Earth, and h is the height of the table. When a body falls, its potential energy is converted into kinetic energy so that the sum of the two energies $E = U + E_K$ remains unchanged. Likewise the tiniest electrically charged particle — an electron — moving across a potential difference of 1 Volt changes its kinetic energy by one electron-volt: 1 eV.

4.5   Momentum in Newtonian mechanics

The term 'momentum' carries many different meanings, of which two are important for us here: 1) a short-duration push setting a body in motion, 2) a vector quantity **p** possessing three components p_x, p_y, p_z and characterizing the moving body. Galileo and Newton were first to understand that a free stand-alone body receiving momentum would in empty space move infinitely long along a straight line at constant velocity. (The fact that the momentum of bodies in our everyday life is not conserved and bodies do not move infinitely long but stop earlier or later is caused by the action of other bodies on them.)

In Newtonian mechanics the momentum of a body is related to its velocity by the formula $\mathbf{p} = m\mathbf{v}$.

The position of a body $\mathbf{r}'$ at time t is connected to its position $\mathbf{r}$ at time $t = 0$ via the Galilean transformation[1]:

$$\mathbf{r}' = \mathbf{r} + \mathbf{v}t \tag{4.2}$$

It is obvious that this transformation stems from the tacit assumption that time t is independent of the reference frame. You will see below that it is very easy to demonstrate that this assumption contradicts the existence of the limiting velocity in nature.

Indeed, if the same initial momentum is imparted to a 10 times lighter body, its velocity would be 10 times higher and it would cover 10 times longer path over the same interval of time.

We shall now repeat the operation of reducing the mass of the particle by a factor of 10. Now its velocity in Newtonian mechanics would have grown 100 times. In Newtonian mechanics, this operation can be performed an unlimited number of times with the same tenfold increase in speed. We recall now that nature has this limiting velocity: $c = 3 \cdot 10^8$ m/s. Then the increase in velocity should begin to decline long before we approach this limit. The increment to kinetic energy must also change. We have to conclude that Newtonian mechanics should be replaced by

---

[1] Here I mean an active transformation when the position of the object undergoes changes, not when the coordinate system does.

some other mechanics. It is called the relativistic mechanics or Einstein's relativity theory. Scientists who successively made key contributions to it were Lorentz, Poincaré, Einstein and Minkowski. Newtonian mechanics (nonrelativistic mechanics) is a special case of the theory of relativity for situations when bodies move at velocities small compared to the speed of light.

4.6 Space and time in relativistic mechanics

To recapitulate: when the velocity of an object approaches the maximum speed c, nonrelativistic Newtonian mechanics becomes completely inapplicable and must be replaced with relativistic mechanics.

Galilean coordinate transformation has been replaced by the Lorentz transformation:

$$\mathbf{r}' = \gamma(\mathbf{r} + \mathbf{v}t)$$
$$t' = \gamma(t + \mathbf{r}\mathbf{v}/c^2), \tag{4.3}$$

where

$$\gamma = 1/\sqrt{1 - v^2/c^2}. \tag{4.4}$$

It is easy to see that as $v/c \to 0$, the Lorentz transformation changes into the Galilean transformation for which $t' = t$. Lorentz transformations implied that not only the spatial coordinates change when the body moves but that the time coordinate changes too. Minkowski, who introduced the concept of four-dimensional space, noticed that

ct and $\mathbf{r}$ become components of one four-dimensional vector. The difference between squared temporal and spatial components gives an invariant quantity known as interval s squared:

$$s^2 = c^2 t^2 - \mathbf{r^2}. \tag{4.5}$$

The four-dimensional space is called, in contrast to the usual three-dimensional space, not Euclidean but pseudo-Euclidean because the invariant here is not the sum but the difference between squares. Note that the difference can be transformed into the sum if instead of t we consider it. As the Minkowski space is isotropic, the magnitude of the four-dimensional interval s, given by formula (4.5) remains unchanged not only under rotations in three spatial planes xy, yz, zx but also in three spatio-temporal planes xt, yt, zt. It is not difficult to see that the last three imaginary rotations correspond to uniform linear motions along the axes x, y, z.

4.7 Energy and momentum in relativistic mechanics

Similarly to this, the isotropic nature of the four-dimensional space implies that energy and momentum of a particle, E/c and $\mathbf{p}$, form a four-dimensional energy-momentum vector, or, as physicists say, the 4-momentum. This 4-momentum squared equals particle's mass squared multiplied by c^2:

$$m^2 c^2 = E^2/c^2 - \mathbf{p}^2 \tag{4.6}$$

As Einstein suggested more than a century ago, in the case of a free body with nonzero mass m this formula reduces to a simple formula

$$E = E_0 + E_K, \qquad (4.7)$$

if we introduce, in addition to the concept of kinetic energy E_K, also the concept of rest energy $E_0 = mc^2$ and the concept of the total energy of a free body E.

The velocity of a free particle is associated with its momentum and energy by the relation

$$\mathbf{p} = \mathbf{v}E/c^2. \qquad (4.8)$$

Or, in a different form,

$$\mathbf{v} = \mathbf{p}c^2/E. \qquad (4.9)$$

Hence as the total energy of a body increases, the increment to its velocity becomes smaller and smaller, and it tends to the limiting value c. This means, however, that Newton's formula eq. (4.1) for kinetic energy becomes inapplicable for $|\mathbf{v}| \approx c$.

The spectacular property of formulas (4.6),(4.8),(4.10) is that they cover not only massive particles but also arbitrarily light particles, and particles whose mass is zero.

4.8 Particle's mass

A different understanding of the mass of a body or particle thus arises. All particles of a given type (e.g. all electrons) are absolutely identical and have the same mass m. For a free particle

$$m^2 = E^2 c^{-4} - \mathbf{p}^2 c^{-2}. \qquad (4.10)$$

4.9   Rest energy

If a free particle is at rest, the rest energy of the particle is given by

$$E(\mathbf{p} = 0) \equiv E_0 = mc^2. \qquad (4.11)$$

If the particle moves slowly, then its energy $E \approx E_0$ and formulas (4.7) and (4.8) imply an approximate equality $\mathbf{p} \approx m\mathbf{v}$, which in Newtonian physics takes the form $\mathbf{p} = m\mathbf{v}$.

4.10   Massless photon

Astronomical observations tell us that the photon mass is very small: $m_\gamma < 10^{-51}$ g. Therefore physicists assume that the mass of the photon is zero. Then equation (4.10) implies that the absolute value of photon's momentum is directly proportional to its energy: $|\mathbf{p}| = E/c$ while equation (4.9) implies that its velocity is always c. Free photons cannot be at rest, they always move at the speed of light.

4.11   Masses of electron and proton

The electron mass m_e is now known to the accuracy of 1 in a hundred million, i.e. 10^{-8} and equals approximately $9 \cdot 10^{-28}$g. The proton mass, known with a comparable accuracy, is approximately two thousand times greater than the electron mass: $m_p \approx 1.7 \cdot 10^{-24}$g.

Chapter 5

Rotation and Quantization

5.1 The spin and orbital rotation*

Two types of particle rotation are known: spin and orbital.

A particle possesses the spin (the intrinsic rotation) regardless of how it moves through space. It is similar to the Earth's rotation around its axis. The spin rotation is characterized by the spin angular momentum $\mathbf{S}$.

Orbital rotation of a particle resembles Earth's rotation around the Sun. It is characterized by the magnitude of the orbital angular momentum $\mathbf{L}$ which is equal to the vector product of the radius vector $\mathbf{r}$ from the center of rotation to the particle by the particle momentum $\mathbf{p}$:

$$\mathbf{L} = \mathbf{r} \times \mathbf{p}, \qquad (5.1)$$

or in vector components:

$$\begin{aligned}
L_x &= yp_z - zp_y, \\
L_y &= zp_x - xp_z, \\
L_z &= xp_y - yp_x.
\end{aligned} \qquad (5.2)$$

It is easy to check that if vector $\mathbf{r}$ is parallel to vector $\mathbf{p}$ then $\mathbf{L} = 0$. In all other configurations $\mathbf{L} \neq 0$. The total angular momentum equals the sum of the orbital and the spin moments: $\mathbf{J} = \mathbf{L} + \mathbf{S}$.

The angular momentum is called a pseudovector or axial vector because, in contrast to ordinary (polar) vectors $\mathbf{r}$ and $\mathbf{p}$, it does not change sign under mirror reflection. For $\mathbf{L}$ this is immediately obvious from the above definition.

The conservation of the angular momentum of an isolated set of particles is implied by the isotropy of space: empty space has no preferred direction.

5.2   About vectors and tensors

In three-dimensional space, the scalar product of two vectors is a scalar, the vector product of two vectors is an antisymmetric tensor (and at the same time — axial vector) having three components. It is easy to see that a symmetric tensor constructed of two vectors has five (not six) components as it does not contain the scalar: $3 \times 3 = 1 + 3 + 5$.

5.3   The orbital angular momentum in theory of relativity

In relativity theory the three components of (5.1) are a part of the four-dimensional tensor whose other three components have the form:

$$\begin{aligned}
N_x &= ctp_x - xE/c, \\
N_y &= ctp_y - yE/c, \\
N_z &= ctp_z - zE/c,
\end{aligned} \tag{5.3}$$

or in vector form

$$\mathbf{N} = ct\mathbf{p} - \mathbf{r}E/c. \tag{5.4}$$

If the vectors $\mathbf{r}$ and $\mathbf{p}$ are not parallel to each other, then $\mathbf{N} \neq 0$. If the vectors $\mathbf{r}$ and $\mathbf{p}$ are parallel to each other, than $\mathbf{N} = 0$ and

$$ct\mathbf{p} = \mathbf{r}E/c. \tag{5.5}$$

In the case of translational motion $\mathbf{r} = \mathbf{v}t$ and we derive equations (4.8) and (4.9).

5.4 Identity of particles*

All elementary particles of a given type (e.g. all electrons) are absolutely identical. (Among other things, their masses are exactly the same.) Hence all atoms of a given sort are identical.

5.5 Quantization of S and L*

Another remarkable property of nature is that rotation is quantized. The components of $\mathbf{L}$ and $\mathbf{S}$ along some external axis (say z) assume only discrete values with an increment of $\hbar$ where $\hbar$ is the quantum of action introduced into physics in the early twentieth century by Planck.[1]

In conditions when quantization is absolutely essential, the motion and interaction of particles is described not by Newtonian mechanics but by quantum mechanics. This is true not only for the theory of elementary particles, theory of atoms and theory of atomic nuclei but also for the theory of molecules and the theory of condensed state.

---

[1] In 1900 Planck introduced the quantity equal to $2\pi\hbar$ into the theory of black body radiation.

According to quantum mechanics the square of the the orbital angular momentum **L** has to satisfy the equation

$$\mathbf{L}^2 = \hbar^2 l(l+1), \tag{5.6}$$

where $l = 0, 1, 2, \ldots$ is the orbital quantum number. To each value of l there correspond $2l+1$ projections on a certain axis in space: $1, 3, 5 \ldots$. See section 5.2 above.

5.6 More about spin*

A spectacular property of elementary particles is that they have spin — their intrinsic (proper) rotation. I would be inclined to compare them with tiniest tops. However, an ordinary top may not be rotating while the rotation of an elementary particle with nonzero spin cannot stop. The spin differs from the angular momentum of orbital motion in that an absolutely free particle not executing orbital rotation has the spin nevertheless. As will be explained below, the particle spin plays key role in the formation of systems containing many particles.

5.7 Fermions and bosons*

All particles fall into one of two large classes: fermions (with half-integer spin) and bosons (with integer spin). The electron whose spin is $\frac{1}{2}$ is a fermion. The photon whose spin is 1 is a boson. System consisting of two or more identical particles radically differ from each other depending on whether these particles are fermions or bosons.

Not more than one fermion can occupy a given quantum state. Any number of bosons can occupy a given quantum state.

5.8   Elementary quantum state*

Another very important concept in quantum mechanics is that of the elementary quantum state. The elementary quantum state of a free particle is completely characterized by its momentum and orientation of its spin. The elementary quantum state of a bound particle is completely characterized by several quantum numbers (see below). The above-formulated statement on the identity of atoms assumes that atoms occupy identical states. In what follows I will, for the sake of brevity, simply use the term "state" although it would perhaps be wiser to replace the words "elementary quantum state" by some other special term: "state" in quantum mechanics has nearly a dozen different interpretations.

5.9   Bound states*

A bound state of two particles with masses m_1 and m_2 is characterized by the mass of this state $m = m_1 + m_2 - \varepsilon/c^2$ where ε is the binding energy. The quantity ε/c^2 is called the mass defect. In the case of a bound state of an arbitrary number of particles, ε is the total amount of energy required for making every particle free from every other particle of the system.

Chapter 6

Particles as Corpuscles and Waves

6.1 Wave vector*

The constant $\hbar$ plays a most important role not only in rotation of elementary particles but also in their entire behavior. The thing is that unlike ordinary particles (corpuscles) of classical mechanics, elementary particles are some kind of "centaurs", so to speak, possessing the properties of corpuscles and those of waves at the same time.

The state of a free particle is characterized in quantum mechanics by a certain value of the wave vector $\mathbf{k}$ related to the particle's momentum $\mathbf{p}$ via the de Broglie relation:

$$\mathbf{p} = \hbar\mathbf{k}. \tag{6.1}$$

The quantity $k = |\mathbf{k}|$ is known as the wave number. It is related to the wavelength λ by the formula

$$k = 2\pi/\lambda. \tag{6.2}$$

In the limit in which the wavelength becomes negligibly small, we return to classical mechanics or geometrical optics. (Of course, we mean phenomena in which interference is nonessential.)

A formula similar to (6.1) relates the energy of a particle E with the frequency ω characterizing it (its wave):

$$E = \hbar\omega. \tag{6.3}$$

6.2 The wave function*

As a rule, an introduction into quantum mechanics begins not with the concept of quantum state but with that of the wave function: $\psi(\mathbf{r}, t)$ where $|\psi(\mathbf{r}, t)|^2$ is the probability that the particle has a particular value of $\mathbf{r}$ at the moment of time t. Unfortunately, many people form a very false impression here that in contrast to classical mechanics, in quantum mechanics everything has probabilistic nature. In reality, though, quantum mechanics is in many ways incomparably more rigid than the classical, and we will have a chance to observe it. This inflexibility manifests itself most strikingly in the structure of the Periodic table of chemical elements; to understand it, one only requires the concept of quantum state but does not need to resort to the concept of probability. For this reason, reading sections 6.8–6.10 of this chapter (which deals with wave equations) is not necessary for understanding the structure of the Periodic table.

6.3 Probability amplitude*

The quantum-mechanical probability amplitudes were invariably calculated for more than twenty years by solving wave equations (we shall look at wave equations in more detail later in this chapter). With this approach, one needs to introduce such concepts as the Hilbert space and the operators acting in it. The majority of textbooks still teach quantum mechanics in this way.

This book attempts not to use these concepts, using instead diagrams invented by Richard Feynman in mid-20th century for calculation of probability amplitudes. These diagrams will be introduced in the chapter devoted to quantum electrodynamics, the QED. They are widely used for caclulating probabilities of various processes.

6.4   The role of chance in the decay*

Processes of decay of elementary particles and atomic nuclei and processes of emission of photons by atoms are characterized by the lifetime τ; of course, lifetimes are different for different processes. The number of particles (or nuclei, or atomic levels) non-decayed over a time τ decreases by a factor of $e \approx 2.7$. (Decay half-life $\tau_{1/2}$, over which the number of non-decayed particles is reduced by half is often used instead of lifetime; $\tau_{1/2} = \tau \ln 2 \approx 0.7\tau$.) Quantum mechanics and relativity theory allow us to predict with a high degree of accuracy the value of τ; however, we cannot predict in principle when a specific particle will decay. Here chance rules; random phenomena are described by the probability theory.

6.5 The role of chance in the two-slit experiments

The confirmation of wave properties of microscopic particles passing through two slits of the first screen and creating an interference pattern on the second screen standing behind the first played an important role in the evolution of quantum mechanics. The pattern is the sharper, the greater the number of particles passing through the slits. Note that particles can pass through the slits one-by-one; this proves that an individual particle possesses wave properties. However, it is absolutely impossible to predict at what point the particle hits the second screen. Here again chance reigns. Many authors writing about quantum mechanics do not appreciate this behavior and attempt to apply the classical concept of trajectory. Thus a "theoretical interpretation" was produced some time ago of quantum mechanics as a science that claims to describe the coexistence of many different worlds; it continues to cause debates.

6.6 Uncertainty relations*

Heisenberg's uncertainty relations play a very important role in quantum mechanics:

$$\begin{aligned}
\Delta E \Delta t &\geq \hbar/2, \\
\Delta p_x \Delta x &\geq \hbar/2, \\
\Delta p_y \Delta y &\geq \hbar/2, \\
\Delta p_z \Delta z &\geq \hbar/2.
\end{aligned} \tag{6.4}$$

These relations establish a connection between the uncertainties with which a particle's energy and the time of its observation, or its momentum or its coordinate can be measured.

6.7   "Correct" and "incorrect" questions*

Quantum mechanics is incomparably more strict than classical mechanics when it answers "correct" questions. But to "incorrect" questions it offers only probabilistic answers. An example of correct question: "What quantum numbers does the particle have in a given state?". Examples of incorrect questions (incorrect by virtue of the uncertainty relation): "Where a free particle with a specific momentum is located?" or "What is the trajectory of a particle?". As a result of this, such very important concept of classical physics as location in the world (in four-dimensional spacetime) becomes probabilistic ("smeared").

6.8   Schrödinger equation

In the nonrelativistic limit the wave function satisfies the Schrödinger wave equation:

$$i\frac{\partial \psi}{\partial t} = \hat{H}\psi, \qquad (6.5)$$

where $\hat{H}$ is the Hamiltonean operator known as the Hamiltonian; it equals the sum of the kinetic and potential energy operators. For a charged particle in electromagnetic field $\phi, \mathbf{A}$ it has the form:

$$\hat{H} = \frac{1}{2m}(\hat{\mathbf{p}} - e\hat{\mathbf{A}})^2 - e\phi, \qquad (6.6)$$

where $\hat{\mathbf{p}} = -i/\partial\mathbf{r}$ is the momentum operator. The wave function ψ is the set (linear superposition) of all possible quantum states of a given physical system.

6.9   The Klein–Fock–Gordon equation

The equation for a relativistic spinless particle was found in 1926 by Klein, Fock and Gordon. For a free particle it has the form

$$(\frac{\partial^2}{\partial t^2} - \frac{\partial^2}{\partial x^2} - \frac{\partial^2}{\partial y^2} - \frac{\partial^2}{\partial z^2} + m^2)\psi = 0. \qquad (6.7)$$

In four-dimensional notation, it can be written as

$$(\frac{\partial}{\partial x^{\mu}}\frac{\partial}{\partial x_{\mu}} + m^2)\psi = 0. \qquad (6.8)$$

Here x^{μ} is a contravariant four-dimensional vector and x_{μ} is a covariant four-dimensional vector. Repeated ("dummy") indices imply summation: $x_{\mu} = g_{\mu\nu}x^{\nu}$.

The relativistically invariant wave function of a spinless charged particle in electromagnetic field $A_{\mu} = \phi, \mathbf{A}$ satisfies the equation

$$((\frac{\partial\psi}{\partial x^{\mu}} - ieA_{\mu})(\frac{\partial\psi}{\partial x_{\mu}} - ieA^{\mu}) + m^2)\psi = 0. \qquad (6.9)$$

6.10 Dirac equation

The relativistic wave function for a free particle with spin $\frac{1}{2}$ satisfies the Dirac equation:

$$(\hat{p} - m)\psi = 0, \tag{6.10}$$

where $\hat{p} = p_\mu \gamma^\mu = i\frac{\partial}{\partial x^\mu}\gamma^\mu$ while γ^μ are four Dirac matrices of dimension 4×4:

$$
\gamma^0 = \begin{pmatrix} 1 & 0 & 0 & 0 \\ 0 & 1 & 0 & 0 \\ 0 & 0 & -1 & 0 \\ 0 & 0 & 0 & -1 \end{pmatrix} \quad
\gamma^1 = \begin{pmatrix} 0 & 0 & 0 & 1 \\ 0 & 0 & 1 & 0 \\ 0 & -1 & 0 & 0 \\ -1 & 0 & 0 & 0 \end{pmatrix},
$$

$$
\gamma^2 = \begin{pmatrix} 0 & 0 & 0 & -i \\ 0 & 0 & i & 0 \\ 0 & i & 0 & 0 \\ -i & 0 & 0 & 0 \end{pmatrix}, \quad
\gamma^3 = \begin{pmatrix} 0 & 0 & 1 & 0 \\ 0 & 0 & 0 & -1 \\ -1 & 0 & 0 & 0 \\ 0 & 1 & 0 & 0 \end{pmatrix}. \tag{6.11}
$$

These Dirac matrices can be compactly expressed in terms of 2×2 matrices one of which is a three-dimensional scalar:

$$\mathbf{1} = \begin{pmatrix} 1 & 0 \\ 0 & 1 \end{pmatrix},$$

and the other three matrices, known as Pauli matrices, are components of the three-dimensional vector $\boldsymbol{\sigma}$:

$$\sigma^1 = \begin{pmatrix} 0 & 1 \\ 1 & 0 \end{pmatrix}, \quad \sigma^2 = \begin{pmatrix} 0 & -i \\ i & 0 \end{pmatrix}, \quad \sigma^3 = \begin{pmatrix} 1 & 0 \\ 0 & -1 \end{pmatrix}. \tag{6.12}$$

So we obtain

$$\gamma^0 = \begin{pmatrix} \mathbf{1} & 0 \\ 0 & -\mathbf{1} \end{pmatrix}, \quad \boldsymbol{\gamma} = \begin{pmatrix} 0 & \boldsymbol{\sigma} \\ -\boldsymbol{\sigma} & 0 \end{pmatrix}. \tag{6.13}$$

In what follows (in Chapter 20) we will also use the matrix $\gamma^5 = -i\gamma^0\gamma^1\gamma^2\gamma^3$, which can be presented in the form

$$\gamma^5 = -\begin{pmatrix} 0 & 1 \\ 1 & 0 \end{pmatrix}. \tag{6.14}$$

6.11 Action

Among all physical quantities there is one which occupies a central position in physics. This quantity is action S. The central role of action in physics is due to the existence of the basic law of physics — the principle of least action. This principle was first formulated by Fermat. The universal, the key role of action in physics became clear only in the 20th century.

In the simplest case the action of a free particle that covered in time t the distance from the origin to the point $\mathbf{r}$ is defined as

$$S = -Et + \mathbf{pr}. \tag{6.15}$$

If a free particle is at rest, the action for it is $S = -E_0 t = -mc^2 t$.

The solution of the wave equation can be written as

$$e^{iS/\hbar}. \tag{6.16}$$

Further discussion of S would stretch beyond the confines of this small book. A beautiful explanation of the physical meaning of the least action principle see in Feynman Lectures [5].

Chapter 7

More About Units*

7.1 Units: experiment and theory

Talking about units here may seem out of place but it is necessary because physics is a science not only theoretical but also — and first of all — experimental: theory organizes facts that were established experimentally. According to experiments, $\hbar = 1.054\,571\,628(53) \cdot 10^{-34}$ J·s where 1 J stands for one joule and 1 s is one second; here and further down the digits in parentheses ((53) in this case) give the experimental uncertainty in the last significant digits.

7.2 About SI system of units

The first international agreement to establish the SI system of units was signed in the 19th century in response to the needs of trade, technology and science. The unit of energy in SI, one Joule, equals 1 kg·m^2·s^{-2}. Another definition of Joule in SI in terms of Coulomb and Volt: 1 J = 1 C·1 V. Obviously, neither kilogram nor Joule are convenient units when discussing elementary particles.

7.3   Electron-volt

In experimental elementary particles physics, the generally used unit of energy is the energy that an electron
gains having crossed the potential difference of one volt.
This unit is called one electron-volt — 1 eV. Also used are
derivatives of this unit: milli-, kilo-, mega-, giga-, tera-eV
(1 meV= 10^{-3} eV, 1 keV= 10^3 eV, 1 MeV= 10^6 eV, 1
GeV= 10^9 eV, 1 TeV= 10^{12} eV). As the electron charge
equals $1.6 \cdot 10^{-19}$ C, we find that 1 J = $6.2 \cdot 10^{18}$ eV. Hence
$\hbar = 6.582\,118\,99(16) \cdot 10^{-16}$eV·s.

7.4   Units in which $c, \hbar = 1$

The system of units most often used in elementary particle
physics chooses the speed of light c for the unit of speed
and $\hbar$ for the unit of angular momentum. This greatly
simplifies all equations as we then can and should assume
$c, \hbar = 1$ in each. In units where $c = 1$, energy, momentum
and mass have one and the same dimension. Thus the mass
of particles is given in electron-volts: 1 eV = $1.78 \cdot 10^{-36}$
kg, $m_e = 0.51$ MeV, $m_p = 938$ MeV ≈ 0.9 GeV. Likewise,
r and t also have the same dimension.

7.5   On choosing the system of units

The choice of the system of units needed in the solution of
a specific physical problem is a matter of convenience. The
system of units $c, \hbar = 1$ is the most useful when considering the fundamental issues in relativistic quantum physics.

Anyone who wants to understand modern physics needs to mastering it. This does not imply, however, that this system is always convenient to work with, e.g. in the classical or non-relativistic limits.

Chapter 8

The Hydrogen Atom*

8.1   On potential energy

We know that the lightest of atoms — the hydrogen atom — consists of a proton and an electron. The mass of a hydrogen atom is mainly determined by the mass of the proton m_p but all other properties are dictated by the mass of the electron m_e and its electric charge $-e$ which is equal in magnitude and opposite in sign to the charge of the proton $+e$. The potential energy of attraction between the electron and the proton at a distance r from one another equals

$$U = -e^2/r = -4\pi\alpha\hbar c/r, \qquad (8.1)$$

The key role in the theory of the atom is played by the dimensionless quantity $\alpha = e^2/4\pi\hbar c \approx 1/137$. Note that the potential energy of the electron but is independent of its velocity. The potential energy can be introduced only in the non-relativistic approximation. (I remind the reader that $e = 1,6 \cdot 10^{-19}$ C (see 7.3) but we do not use the SI system here (see 12.4).)

8.2   Electron–proton interaction

The mechanism of interaction between electrons and protons consists in their exchanging photons with one another. (This will be explained in more detail later in the book.) Two situations need to be distinguished here: the elastic and inelastic ones. In the former case two free particles approach each other, their initial momenta are changed by their interaction, and they fly away never to meet again unless the surrounding particles intervene. In the latter case an electron and a proton form a bound system and cannot fly apart. The first type of motion is known as infinite, the second type — finite since the motion occurs in a bounded part of space (For an elastic, free situation to convert to an inelastic, bound one it is sufficient for the electron to emit a photon carrying away the excess energy. The reverse process in which a bound electron absorbs a photon and becomes free is referred to as ionization of the atom.)

8.3   Principal quantum number

The behavior of a bound electron–proton is dictated by the laws of quantum mechanics. Their binding energy ε can assume only specific (quantized) values. In the lowest approximation in α the binding energy has the form:

$$\varepsilon_n = m_e c^2 \alpha^2 / 2n^2, \qquad (8.2)$$

where n is the so-called principal quantum number which can take only integer values from 1 to infinity. The greater

n, the smaller the binding energy of the atom and the easier it is to break this bond. The binding energy of the electron in the ground state of the hydrogen atom is $\varepsilon_1 = m_e c^2 \alpha^2 / 2 = 13.6$ eV. It is known as the Rydberg energy.

8.4 Mass of quantum state

I need to emphasize that a quantum state has no specific value of kinetic energy, nor a specific value of potential energy. It only has a definite value of the total rest energy of the given state or of its mass. Thus the mass of the ground state of the hydrogen atom is $m_H = m_p + m_e - \varepsilon_1 / c^2$.

8.5 Orbital quantum number

The state of the electron depends, in addition to the principal quantum number n, also on the orbital quantum number l, and such that $l \leq n-1$. The orbital quantum number describes the orbital angular momentum of the electron. I need to emphasize that the notion that the electron moves along its orbit as, say, a planet does is valid (though only approximately) only at very large values of $l \gg 1$. At $l = 0$ an electron resembles a spherically symmetrical cloud rather than a classical orbiting particle. Concepts such as the location in space $\mathbf{r}$, momentum $\mathbf{p}$ and velocity $\mathbf{v}$ have no meaning for it. Note also that the mass of the state in the hydrogen atom in the lowest approximation in α

depends only on the principal quantum number n and is independent of all other quantum numbers (including also l).

8.6   The projections of L and S

The state of the electron in the atom is described, in addition to n and l, by two more quantum numbers which characterize the projection of the orbital angular momentum and spin of the electron onto some spatial axis z. The projection of the orbital quantum number can assume $2l+1$ values: $l_z = l, l-1, l-2, \cdots -l+2, -l+1, -l$; and the spin projection can assume two values: $s_z = +\frac{1}{2}, -\frac{1}{2}$. Usually the direction of the axis z is imposed by the external magnetic field in which the atom was placed. (The quantity l_z is very often denoted by the combination m_l.)

8.7   The emission and absorption of light

The transition of an atom from the more excited (higher) state to a less excited one or to the ground state is accompanied with the emission of a photon while the reverse transition is a result of absorption of a photon.

Chapter 9

Periodic Table of Chemical Elements

9.1 From protons to nucleons*

Even elementary knowledge of quantum mechanics makes it possible to understand the principal structure of the Mendeleev Table without difficulties.

We cannot understand the structure of atoms heavier than hydrogen in terms of only three particles (e, γ, p) since their nuclei contain neutrons as well as protons. The neutron is an electrically neutral particle with mass very close to that of the proton: $m_n - m_p \approx 1.29$ MeV. The presence of neutrons in atomic nuclei is essential for nuclear physics but much less so for atomic physics. Indeed, the chemical properties of isotopes — those elements which contain the same number of protons in the nuclei but a different number of neutrons — are very similar.

In those situations in which the differences between protons and neutrons are insignificant they are given the common name nucleons. The total number of nucleons in a nucleus is denoted by the letter A. It is A that determines the mass of the nucleus. The total number of protons in the nucleus is denoted by the letter Z. It is Z

that determines the charge of the nucleus. Since atoms are electrically neutral the number of electrons in an atom is also equal to Z.

9.2   Pauli exclusion principle. Fermions and bosons*

A key role in the structure of atoms is the Pauli exclusion principle. As already mentioned earlier, the electron spin equals $\frac{1}{2}$. Electrons obey the Pauli principle which states that not more than one electron can occupy a given quantum state. A similar principle is true for all particles with half-integer spin (they are called fermions). I have to add that particles with integer spin are called bosons. All particles of a given type (e.g. all electrons or all photons) are exactly identical. Not more than one fermion (electron) can occupy a given state; by contrast, there can be any number of bosons (photons) occupying the same state.

9.3   Horizontal periods of the periodic table of elements*

In order to understand how the periodic table is arranged (see, e.g. [6]), one needs to remember that the state of the electron in an atom is described by four quantum numbers: 1) prinipal n, 2) orbital l, 3) its projection m_l, 4) spin projection m_s. The principal quantum number n can assume integer values 1, 2, 3,.... The orbital number l can take integer values from 0 to $n - 1$. States with $l = 0$, 1, 2, 3 are denoted by s, p, d, f, respectively. The projection m_l can assume $2l + 1$ values from $-l$ to $+l$. The projection of the spin m_s can assume two values: $-\frac{1}{2}$, $+\frac{1}{2}$. In contrast to the hydrogen atom, the mass of the quantum state of other atoms depends not only on n but also on l. It is for this reason that d shells in the fourth and fifth periods are filled before the p shells and in the sixth and seventh periods the f shells are filled before the d shells.

Note that the s, p, d, f shells were assigned names long before the notions of atoms and electrons were introduced. In the 19th century these letters denoted the terms of optical spectra of various chemical elements. The frequency of any spectral line can be found as a difference between frequencies of two terms. This may be a good place to mention that helium became known in 1868 as a result of studying the spectra of solar prominences.

9.4 First period

The first period of the periodic system (which has only one electron shell, $1s$) consists of atoms of two elements: hydrogen ^{1}H ($1s^1$) and helium ^{2}He ($1s^2$). Here and further on the upper left-hand index in the notation of the atom indicates the total number of atomic electrons in this atom and an equal number of protons Z in the nucleus of the atom; the number placed in front of the symbol of the shell with the l indicates the value of its principal quantum number n; index at the upper right index in the shell notation indicates how many electrons are located on this shell.

9.5   The second and third periods

The second period (its elements have the inner electron shell $1s$ and the outer shell $2s2p$) consists of atoms of eight elements: from lithium ^{3}Li $(2s^1)$ to neon ^{10}Ne $(2s^22p^6)$. The reader remembers that the number of electrons on any filled s shell is two, and on any filled p shell it is six. The third period (having the outer shell $3s^23p^6$) also consists of atoms of eight elements: from sodium ^{11}Na to argon ^{18}Ar.

9.6   The fourth and fifth periods

The fourth period (having the outer shell $4s3d4p$) consists of atoms of eighteen elements: from potassium ^{19}K $(4s^1)$ to krypton ^{36}Kr $(4s^23d^{10}4p^6)$. The fifth period (having the outer shell $5s4d5p$) also consists of atoms of eighteen elements: from rubidium ^{37}Rb $(5s^1)$ to xenon ^{54}Xe $(5s^24d^{10}5p^6)$.

9.7   The sixth and seventh periods

The sixth period (having the outer shell $6s4f5d6p$) consists of atoms of thirty-two elements: from cesium ^{55}Cs $(6s^1)$ to radon ^{86}Rn $(6s^24f^{14}5d^{10}6p^6)$. The seventh period (having the outer shell $7s5f6d7p$) also consists of atoms of thirty-two elements: from francium ^{87}Fr $(7s^1)$ to the element No

118 $(7s^2 5f^{14} 6d^{10} 7p^6)$. Elements 113–118 have not been named so far.

9.8 8 groups and 18 vertical columns of the table

18 vertical columns of the Periodic Table are usually numbered with Arabic numerals. In fact in chemistry these same columns are denoted by Roman numerals from I to VIII, followed by letters A or B (see [6]); columns with the same Roman numerals form what is known as groups. (So that the group VIIIB consists of columns 8,9,10.) The first column (IA) of the periodic table of elements is the home of the atoms of hydrogen and alkaline metals; they all have only one electron in the outer shell. The second column (IIA) is occupied by alkaline-earth elements with two electrons on the outer shell. The columns (IIIA), (IVA), (VA), (VIA) begin with boron, carbon, nitrogen and oxygen, respectively. The column (VIIA) consists of chemically active elements which lack one electron in the outer shell. And finally the last (eighteenth) column (VIIIA) is occupied by the atoms of noble gases with completely filled outer shell.

The columns 2–17 of the first period are empty. The columns 3–12 of the second and third period are empty too.

The rows 3–12 of the fourth and fifth periods are filled with transition metals (these rows are denoted by IIIB, IVB, VB, VIB, VIIB, VIIIB, IB, IIB, respectively).

The sixth and seventh period have in the third column not one but 15 elements: these have identical outer shells sd but differently filled lower-lying shells f. These elements are chemical analogues of lanthanum — lanthanides (from ^{57}La to ^{71}Lu) and correspondingly of actinium — actinides (from ^{89}Ac to ^{103}Lr). This is why they have not 18 but 32 elements each: $32 = 18 + (14 = 15 - 1)$.

Chapter 10

Substance

10.1 Molecules*

As atoms approach one another, they exchange electrons and bind together, forming a molecule. The number of atoms in a molecule ranges from two or three (H_2, O_2, HCl, H_2O, CO_2) to hundreds of millions (as in the case of the molecule of life — the DNA — deoxyribonucleic acid). By the order of magnitude the binding energy of atoms in molecules is about one electron-volt.

10.2 Gases*

We find individual molecules in gases around us: for example, air contains molecules of O_2, N_2, CO_2, H_2O. The molecules of noble gases He, Ne, Ar,... are monatomic. However, in most phenomena occurring at normal temperatures and pressures we observe condensed substances in the form of solids and liquids. Transitions between different states of matter caused by changes in temperature and pressure are called phase transitions. Melting and evaporation of water are examples of such transitions.

10.3   Loschmidt number*

At normal pressure and temperature air molecules are
moving with an average energy on the order of eV/30.
The distance between molecules exceeds their diameters
by about an order of magnitude. One cubic meter of air
at normal conditions contains approximately $2.7 \cdot 10^{25}$ gas
molecules. This is the Loschmidt number [7].

10.4   Temperature*

Temperature characterizes the mean energy of particles of
matter. It is typically measured in degrees Celsius ($^\circ$C)
and denoted by t. The freezing temperature of water is
chosen for the origin $t = 0^\circ$C and its boiling temperature
for 100°C. The concept of temperature is best discussed
by considering as an example the ideal monatomic gas in
which the interaction between particles is assumed absent.
In this case the total energy reduces to the kinetic energy
of particles: $\bar{E} = \frac{3}{2}kT$ where $\bar{E}$ is the mean particle energy,
T is the absolute temperature ($T = t + 273$) in Kelvin (K),
and k is the Boltzmann constant: $k = 8.62 \cdot 10^{-5}$ eV/K. In
contrast to the universal constants c and $\hbar$, the constant k
is not more than a factor for conversion from temperature
units to energy units.

10.5   More on universal constants

This is a good place for re-emphasizing the exceptional role
played by the universal (world) fundamental constants.

Obviously, c and $\hbar$, as well as k are conversion factors. (For instance, one can use c to convert light years into kilometers.) However, the transition from the theory of relativity to Newtonian mechanics occurs only when c tends to infinity, and the transition from quantum to classical mechanics occurs only when $\hbar$ tends to zero. Nothing of the sort holds for k which relates the mean energy of a system of many particles to the temperature of this system. (If energy and temperature are measured in the same units then k is not included in the definition of entropy either.) This is the reason why the Boltzmann constant, so important in thermodynamics and statistical physics is not a fundamental and universal (world) constant.

10.6   Condensed matter*

Atoms of condensed substances are densely packed. They touch one another. Molecules of the liquid phase can move and swap places. Ideally, the positions of molecules (or atoms) of the solid phase are strictly fixed though in fact these fixed points are the averaged positions of the oscillating atoms and molecules. Solids can be classified in a number of ways, for example, into amorphous bodies and crystals. An example of the former is glass; it consists of molecules of SiO_2. An example of the latter is common salt $NaCl$. Typical examples of crystaline structure are metals although exceptions from this are known, e.g. mercury Hg which is liquid under normal conditions.

Another classification of solids is their division into conductors, dielectrics and semiconductors. Conductors are so called because they conduct electric currents well and are used to fabricate wires; dielectrics are poor conductors and materials for making insulators. The conductivity of semiconductors depends on external conditions. The entire electronics is based on semiconductors.

10.7   Crystallization*

Crystallization is a process in which atoms self-organize, settling into very specific positions in space and creating spatial lattices whose structure is dictated by the chemical properties of atoms: their valence, i.e. their ability to attract electrons from neighboring atoms, or to share their electrons with them. The result is the creation of a steady state of a large (macroscopic) number of atoms with minimum total energy. What emerges is referred to as long-range order.

10.8   Phase transitions*

Mutual transformations of various phase states in response to changes in temperature and/or pressure are known as phase transitions. The hottest phase transitions: melting of tungsten and its boiling. Here are some cold phase transitions: formation of solid hydrogen at $T = 14\,\mathrm{K}$, transition of liquid ${}^4\mathrm{He}$ ($T = 2.17\,\mathrm{K}$) and liquid ${}^3\mathrm{He}$ ($\mathrm{T} \approx 10^{-3}\,\mathrm{K}$) to superfluid state.

10.9 Superfluidity and superconductivity

Superconductivity was first observed in 1911 in mercury wire. Its electrical resistance would vanish at a temperature below 4 K. Superfluidity was discovered in 1938 in liquid helium. At temperatures below 2 K this liquid flows without friction.

Hundreds of substances displaying superconductivity have been discovered in the last 100 years. First high-temperature alloys which became superconducting at temperatures above 30 K were discovered in 1986. Since then the upper limit of high-temperature superconductivity has surpassed 125 K. The theoretical model of superconductivity was created in 1957. It was based on the concept of superfluidity of the so-called Cooper pairs formed by two electrons with equal in magnitude and oppositely directed momenta and antiparallel spins. (Electrons are attracted to each other due to their interaction with the excitations of the crystalline lattice of the solid.)

10.10 Quasiparticles

Very useful in the quantum many-particle theory is the concept of quasiparticle which stands for collective quantum excitation of substance with a definite energy, momentum and spin. Examples of quasiparticles are phonons, polarons and excitons in crystal lattices, magnons in spin systems of magnets, phonons and rotons in liquid helium and plasmons in plasmas. The concept of quasiparticle makes it possible to reduce complex problems of interaction between particles to a simpler analysis of a gas of quasiparticles.

Chapter 11

Quantum Electrodynamics — QED

All the variety of phenomena in Chapters 9 and 10 arises as a consequence of the interaction between virtually two particles: the photon and the electron as in the above-discussed phenomena nuclei do play the role of the rather passive architectural stage set. A complete theory of the photon–electron interaction is called Quantum Electrodynamics (QED).

11.1 QED from Dirac to Feynman*

QED was created by the efforts of many physicists. However, it is impossible not to mention the names of two of them. Paul Dirac laid the foundations for QED at the end of the 1920s, and Richard Feynman in the late 1940s gave it that very simple form in which it can now be explained to beginners.

11.2 Lamb shift

The measurement at the end of the 1940s of the so-called Lamb shift of the $2s$ and $2p$ levels of the hydrogen atom was a most important experimental discovery. As mentioned in Chapter 8, these levels have the same energy in the lowest approximation in α: both lie above the level $1s$ by the amount $3\varepsilon_1/4$. Experiments showed though that $2s$ lies above $2p$ by approximately two millionth of an electron-volt and can decay to it with emission of a microwave photon (at frequency 1058 Megahertz). It was soon calculated in the framework of QED that this splitting is contributed by higher approximations of perturbations theory in α and equals $0.82\alpha^3\varepsilon_1$.

11.3 Positron and other antiparticles*

A most important factor in QED is the fact that the electron has its antiparticle — the positron with exactly the same mass and the same absolute value of the electric charge, but the sign of the charge is positive in contrast to the negative electron. All charged particles (hence fermions and bosons) and many of electrically neutral particles such as neutron, have their antiparticles. The neutron and the antineutron are different particles. Some particles, however, absolutely coincide with their antiparticles. They are known as genuinely neutral. The photon is an example of a genuinely neutral particle.

11.4 Feynman diagrams*

Consider now the scattering of an electron by a proton occurring through the exchange of a photon. Arrows on the figure indicate the direction of the time flow for each particle. The initial state is on the left and the final is on the right. The 4-momenta of each of the particles change as a result of interaction.

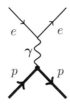

Figure 11.1   Scattering of electrons by protons

Electron and proton lines correspond to free particles; for them squared 4-momentum p equals mass squared: $p^2 \equiv E^2 - \mathbf{p}^2 = m^2$. (Here we use the system of units in which $c = 1$.) We say about these particles that they are on-shell and that they are real. Real particles either arrive from infinitely large distances or fly away infinitely far.

The photon line in Fig. 11.1 corresponds to a photon with 4-momentum q but it is a virtual, not a real, photon, for which infinity is inaccessible. It is off the mass shell and for this photon $q^2 \neq m_\gamma^2 = 0$. (Note that here and in this and further diagrams straight lines correspond to fermions and wavy lines to bosons.)

Each line and each apex of a Feynman diagram represent a quite specific analytical expression. The diagrams thus act as an unusual "meccano" set designed to calculate quantum amplitudes using "standard set parts"'.

11.5　Backward in time

Let us now consider another process described by the diagram of Fig. 11.2 which is obtained from the previous one by rotating it by 90°. On the left side of this diagram an electron collides with an unusual electron having negative energy which flies backward in time and creates a virtual photon. On the right side a virtual photon creates an ordinary proton and an unusual proton with negative energy which flies backward in time. What physical process is described by this diagram?

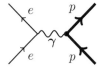

Figure 11.2　Here two of the four particles move backwards in time

Clearly, it describes the birth of a proton–antiproton pair in a collision of an electron and a positron. (This process is possible if the energy of the colliding particles is greater than $2m_p c^2$.)

Figure 11.3   Creation of a proton–antiproton pair in a collision of an electron and a positron

Here e^+ denotes the positron and $\bar{p}$ stands for the antiproton. Such an interpretation of motion back in time was suggested in 1949 by Feynman. It greatly simplified the understanding of physics.

11.6   Antiparticles*

The relativistically invariant description of spinless particles was known since 1926. However, the concept of antiparticle was first introduced only in 1930 for the electron with spin $\frac{1}{2}$. A few years later theorists extended it to the case of bosons. First elementary bosons were discovered by experimenters only in the late 1940s. This is perhaps the reason why in the literature the concept of antiparticle is often tied to the Pauli exclusion principle and to the so-called Dirac sea. The motion backwards in time is equally applicable to fermions and to bosons. It makes unnecessary and insufficiently legitimate the interpretation of positrons as unfilled states in the sea of negative-energy electrons to which Dirac resorted when he introduced the concept of antiparticle in 1930. Indeed, there is no such sea for bosons to which the Pauli exclusion principle does not apply. (It is strange that Feynman chose not to mention this when he said in his Nobel lecture that the Dirac sea is just as good a concept as motion backwards in time.)

11.7   Positronium*

An electron and a positron form an atom resembling the hydrogen atom; it is called positronium. Fig. 11.4 shows photon-induced splitting of positronium into free electron and free positron. (The helical line winding around the electron and positron lines in positronium is a reminder that the two are bound together.)

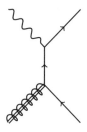

Figure 11.4   Ionization of positronium by a photon

Obviously another diagram can be drawn in which the photon is absorbed not by the electron but by a positron.

11.8   Normal magnetic moment of the electron

The ratio of the magnitude of the magnetic moment of a particle to the quantity $es\hbar/2mc$ where e, s, m are the charge, spin and mass of the particle, is traditionally denoted by the letter g. It is convenient to interpret the vertical outer photon line in the Feynman diagram as representing the photon from the external magnetic field

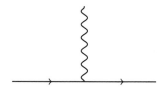

Figure 11.5   The interaction between an electron and the magnetic field

In the lowest order of perturbation theory corresponding to Fig. 11.5, $g = 2$.

11.9   Anomalous magnetic moment of the electron: g-2

Taking into account higher orders of perturbation theory results in $g - 2 \neq 0$. Fig. 11.6 shows the first loop which contributes to $g-2$. At the current stage $g-2$ has been calculated for five loops; the result coincides with the results of accurate measurements.

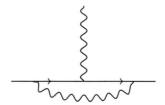

Figure 11.6   Radiative correction to the electron magnetic moment

11.10   Running coupling constant

Consider the phenomenon shown in the following figure in which the photon partly exists as an electron–positron pair. The phenomenon was given the name "vacuum polarization".

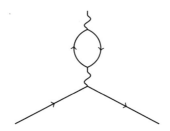

Figure 11.7   Electron–positron loop

Owing to the vacuum polarization virtual electron–positron pairs shield the electric charge of any particle at large distances, i.e. in the cases of small squared momentum of the virtual photon q^2. As q^2 increases, this shielding decreases and the effective charge of the particle increases. A charge which varies with increasing q^2 is called running charge, and the constant $\alpha = e^2/\hbar c$ whose value as a function of q^2 is known as the running coupling constant α. Likewise the electron mass is found to be a running constant because of the loop diagrams.

11.11 Renormalizability of QED

Since α is a dimensionless quantity, all calculations of Feynman loop diagrams in QED can be expressed in terms of the electron mass and the constant α for $q^2 = 0$. This property of the QED is known as its renormalizability.

Chapter 12

Transition to Classical Theory

12.1 Particles or fields?*

We have seen above that the concept of field becomes unnecessary if we switch to the language of Feynman diagrams. The role of fields is now played by virtual particles. Historically, however, the concept of a field preceded those of quantum mechanics and of elementary particles themselves which are treated in quantum field theory as quanta of the field. In my opinion, the "fieldless" Feynman-diagrams approach that I have chosen here is the preferable one for dealing with simple tasks that are solvable by perturbation theory. It will, however, be clear from the subsequent chapters that the quantum field concept has to take over for solving higher-level problems as they make it necessary to go beyond the confines of perturbation theory.

Note that certain features of quantum mechanics that were unacceptable to Einstein and continue to look like paradoxes (such as the familiar "EPR paradox") to many a physicist arise because the ψ-function is treated as a physical field.

12.2 Quasiclassical behavior and the classical limit

Quantum behavior at large quantum numbers is said to be quasiclassical. The shorter the wavelength of the particle compared to the size of the quantum state, the more the state of the particle resembles classical trajectory. Classical concepts are limiting cases of quasiclassical ones. (See section 6.1.)

12.3 The field strength and induction

The strength of the electric field in vacuum **E** is different from the strength of the same field in the medium, i.e. from electric induction **D** in the medium:

$$\mathbf{D} = \epsilon \mathbf{E}, \tag{12.1}$$

where ϵ is the electric permittivity of the medium.

The intensity of magnetic field in vacuum **H** differs from the magnetic induction **B** in the medium:

$$\mathbf{B} = \mu \mathbf{H}, \tag{12.2}$$

where μ is the magnetic permeability of the medium.

By definition, in vacuum $\epsilon_0 = \mu_0 = 1$.

Field propagates in a medium at velocity

$$v = \frac{c}{\sqrt{\epsilon \mu}}. \tag{12.3}$$

12.4 Electric permittivity and magnetic permeability of vacuum*

Concluding the discussion of quantum electrodynamics, it is necessary to say a few words about terminology, metrology and the international SI system of units. The SI system of units has etched into the literature (in fact not only physics literature) the quantities ϵ_0 and $\mu_0 = 1/\epsilon_0 c^2$ known, respectively, as dielectric permittivity and magnetic permeability of the vacuum, these birthmarks of the 19th century. These quantities are closely related to the concept of the ether whose vibrations were then considered to manifest themselves as light waves. The advent of relativity theory made ether redundant; nevertheless, ϵ_0 and μ_0 survived it by a century. These quantities simply do not arise if quantum electrodynamics is consistently presented. This means that ϵ_0 must be set to unity. Also, if $c = 1$ then we have $\epsilon_0 = \mu_0 = 1$. (See e.g. the textbooks [8],[9]). Alas, attempts to persuade metrologists that this should be so meet with their stubborn resistance. Even in the latest issue of the tables of physics constants [10] the constant ϵ_0 enters the definitions of α and the basic characteristics of the hydrogen atom.

The complexity of the SI stems not so much from the diversity of phenomena as from the diversity of theories that describe these phenomena. If everything is reduced to Feynman's mechanics then the diversity of theories evaporates and the system of units becomes simply mechanical. This is very important!

Chapter 13

Gravitation

13.1 Top and bottom*

We get used to the action of the gravitational pull of the Earth on ourselves since our babyhood: climbing upward is hard work, falling downward is easy (albeit not always safe). We do not immediately comprehend that the concepts of the top and the bottom reflect the action of the atoms that make up the Earth on the atoms of our bodies. It takes some time to learn and accept that the Earth is not flat but spherical and what is bottom for us is top for our antipodes.

13.2 The Earth*

A child does not think about the radius of the Earth because the globe is very much bigger than our characteristic dimensions. The radius of the Earth is $6 \cdot 10^3$ km, its mass is $6 \cdot 10^{24}$ kg. The mass of the water envelope is $1.5 \cdot 10^{21}$ kg, hence it is lighter than the Earth by a factor of about 4000. (The dry land area is $1.5 \cdot 10^8$ km^2, the area of the oceans is $3.6 \cdot 10^8$ km^2, the average elevation of the land mass above sea level is 0.9 km, the mean depth of the ocean is 3.9 km.)

The mass of the air envelope (10 m of water equivalent (mwe) ≈ 750 mm Hg) is still smaller by a factor of 400. Both envelopes of the globe are held where they are only by the gravitational attraction of their atoms to those of the solid sphere. This attraction acts constantly both on us and on our antipodes on the opposite side of the globe.

13.3 The inner structure of the Earth

The thickness of the Earth's crust is about 30 km, Its density is about three times that of the density of water. Below the crust lies the twice as dense liquid mantle stretching to a depth of about 3000 km. Located still deeper is the still twice denser solid core.

13.4 Temperature of the Earth

The Earth is very hot inside, so that the average temperature on the surface of the Earth is about 300 K. Temperature at the crust–mantle boundary reaches 700 K, that at the mantle–core boundary — 4500 K, and that at the center of the core is 6400 K.

The high temperature inside the Earth is due partly to the initial high temperature at which it was formed five billion years ago, and partly to the decay of radioactive elements.

(The above numbers are taken from the C. W. Allen's handbook [11], chapter 6: Earth.)

13.5   Tilt of Earth's axis*

Seasonal changes in temperature are caused by the tilt of the Earth's axis of rotation with respect to the plane of the ecliptic (i.e. the plane of Earth's orbit in which the Sun lies) so that summer and winter in the northern and southern hemispheres follow one another. The angle between the equatorial and the ecliptic planes is $23° 27'$ (see chapter 7 of the handbook [11]).

13.6   Newton's Law*

Potential energy of the universal gravitational attraction between two bodies of masses m_1 and m_2 at a distance r from each other is given by Newton's law:

$$U = -G_N m_1 m_2 / r, \qquad (13.1)$$

where Newton's constant $G_N = 6.7 \cdot 10^{-11} \mathrm{m}^3 \, \mathrm{kg}^{-1} \, \mathrm{c}^{-2}$.

13.7   Solar system*

Newton's formula gives the energy of attraction of Newton's apple to the Earth just as it does for the attraction of the Earth to the Sun and of all the other planets of the solar system to the Sun and to one another. The solar mass is $M_\odot = 2 \cdot 10^{30}$ kg; the radius of the Sun $R_\odot = 7 \cdot 10^5$ km. The distance from the Sun to the Earth is $150 \cdot 10^6$ km. Light takes 500 seconds to cover it, or close to eight minutes. This distance is called the astronomical unit (a.u.). The Earth moves in its orbit around the Sun at a velocity of 30 km/s. (This motion prevents Earth from falling onto the Sun.)

The Earth's satellite — the Moon — is 80 times lighter than the Earth, the radius of its orbit is $380 \cdot 10^3$ km.

The biggest and brightest planet in the solar system is Jupiter. It is approximately 300 times heavier than the Earth and the radius of its orbit is 5 times that of the terrestrial orbit.

Not everyone realizes the important role this planet plays in our lives as it diverts the space debris from the asteroid belt towards itself thereby protecting the Earth from the most dangerous collisions.

13.8   The Sun*

The birth and subsequent evolution of the Sun are caused by the gravitational attraction of particles. At the initial stage of contraction of the matter of the gaseous nebula from which the Sun arose, this matter heats up as the gravitational potential energy of particles is converted into their kinetic energy. This transformation of the potential energy into kinetic energy slows down drastically once temperature has reached the level at which nuclear interactions come into play; they convert the rest energy of particles into kinetic energy, thereby preventing the catastrophic gravitational collapse. (On nuclear interactions see below.) Astronomers say that our Sun is some five billion years old: it has lived approximately one half of its lifespan of 10^{10} years. By the end of its life it will successively transform into a red giant as big as the orbit of the Earth, then into a planetary nebula, and finally into a white dwarf of nearly terrestrial diameter.

13.9   Stars in our Galaxy*

Sun is one of the two hundred billion stars that make up our Galaxy — the Milky Way Galaxy. It revolves around the galactic center moving at about 200 km/s at a distance of about 25 thousand light years from the galaxy center. The nearest bright star similar to our Sun is the α Centauri lying at a distance of 4.4 light-years from us (It is visible only in the southern hemisphere). The brightest star in our skies is Sirius lying at a distance of 9 light years. The total number of stars visible to the naked eye is on the order of 700 000. Roughly a third of them lies in the immediate vicinity of the Sun, i.e. in a sphere of a radius of 250 light years. The diameter of the Galaxy is 100 thousand light years, its thickness is 7 thousand light years. The total mass of stars in the Galaxy is $1.4 \cdot 10^{11} M_{\odot}$, the mass of the heaviest stars in the Galaxy is roughly $100 M_{\odot}$; see [11]. Stars whose mass is approximately $M_{\odot}$ live for about 10^{10} years.

13.10   Parsec

Distances to not too distant stars are calculated from the changes in their parallax viewed from different points of the Earth's orbit. One parsec is the distance from which one a.u. is seen as one arc second. This means that 1 ps = $2 \cdot 10^5$ a.u. = $3 \cdot 10^{13}$ km = 3.3 light years. 1 kps equals 3.3 thousand light years. 1 Megaparsec (Mps) = $3.3 \cdot 10^6$ light years. 1 Gigaparsec (Gps) = $3.3 \cdot 10^9$ light years.

13.11 Supernovae

A star essentially heavier than the Sun has essentially shorter life than the Sun. Its death produces a flare of supernova explosion which is brighter than brightness of entire galaxy. This explosion is caused by a gravitational collapse which occurs when inside the star a massive core of non-burning iron is formed. In such explosions elements heavier than iron are created. The last such flare was observed in 1987 and is known as the SN1987A.

The lightest supernovae are created when two stars are close to each other and the heavier of them pulls matter from its lighter partner onto itself. The explosion occurs when it grows heavier than the Sun by a factor of 1.4.

If two stars are close to each other, the heavier of them pulls matter from its lighter partner onto itself. After the former star grows heavier than the Sun by a factor of 1.4 and after the entire hydrogen in its core has been transformed into non-burning iron, gravitation continues to compress the star and in the end produces the so-called supernova flare. The flash produced by the supernova explosion is so bright that at maximum luminosity its brightness is comparable to that of the entire galaxy and may even become brighter.

Such light supernovae with a fixed mass of 1.4 solar masses were used recently as standard "cosmic candles" for establishing the velocities of distant galaxies (see section 15.5).

The explosion of a light supernova transforms the star to a state in which most protons and electrons transform into neutrons and neutrinos (and neutrinos which fly away). This object is known as the "neutron star". A rotating neutron star is a pulsed source of radiowaves — a pulsar. Stars with much larger masses turn into black holes. For more on black holes see section 16.4.

Chapter 14

Other Galaxies

14.1 From our Galaxy to other galaxies*

Lying at a distance of 200 thousand light years are small "satellite galaxies" of out Galaxy — Magellanic Clouds. (By the way, Supernova SN1987A flared up in the Large Magellanic Cloud.) Farther out, 2 million light years away is the nearest to us large Andromeda galaxy which is visible to the naked eye as a tiny fuzzy speck of dust. Astronomers using 20th century telescopes have added 350 billion large galaxies to the list. The total number of stars in all galaxies is on the order of 10^{22}–10^{23}[12].

14.2 Recession of galaxies*

Edwin Hubble discovered in 1929 that the farther the galaxy is from us, the faster it recedes from us: $v = Hr$ where v is the recession velocity, r is the distance to the galaxy and H is the Hubble constant. Later more accurate measurements showed that $H \approx 72$ km/s Mps. The farthest from us galaxies are at a distance $1/H \approx 15$ billion light years. Roughly speaking, distances to galaxies are found from their observed luminosities, and velocities from the red shift of their emission.

14.3   Kinematic shift*

A light source emits a photon with energy E in our direction when it is at rest, but if the same source moves towards us at a velocity v (in units of $c = 1$) then it emits a photon with energy $E\sqrt{\frac{1+v}{1-v}}$, and if it moves away from us at the same velocity then it emits a photon with energy $E\sqrt{\frac{1-v}{1+v}}$. The frequency and wavelength of a photon we observe change correspondingly: as the light source approaches, it light get bluer, and as it recedes, it gets redder. Such a shift in wavelength is known as the kinematic shift.

14.4   Gravitational shift*

We now consider an experiment with a source and a detector of radiation at rest in the static external gravitational field, e.g. in the field of the Earth. This experiment was conducted in the tower of the Harvard University in the 1960s. The source and the detector were placed alternately one in the attic of the tower, the other in the basement. The energy and frequency of a photon moving in a static field does not change but its momentum increases when it flies down, and decreases when it flies up. The upward moving photon gets redder, the downward moving one gets bluer. This shift in wavelength is called gravitational. A

detailed discussion of this phenomenon can be found in
[13]. The statement that static gravitational field changes
the frequency of the photon (encountered often in the literature) is based on the erroneous application of the concept
of potential energy to relativistic particles.

14.5 Quasars and gamma-ray bursts

At distances of billions of light years from us astronomers
found the most powerful sources of radiation in the Universe — quasars and gamma-ray bursters.

The tag "quasar" was formed by merging the words
"quasi" and "star". The emission of radiation by a quasar
may be hundreds of times greater than the normal radiation of a galaxy but its size is relatively very small: it
is comparable with the size of the Solar System. It was
conjectured that the emission of a quasar is produced by
matter falling onto a massive black hole located at the
center of a distant galaxy. Quasars were identified by their
emission in the radio-wave range.

The name "gamma-ray bursters" was given to objects
emitting short bursts of gamma radiation lasting for seconds or minutes.

Chapter 15

Big Bang

15.1 The expanding Universe*

A big explosion (The "Big Bang") occurred approximately 15 billion years ago, which gave rise to our Universe. In its first moments it expanded exponentially. This was the so-called inflation that was responsible for the uniformity and isotropy of the Universe. (Much later very small nonuniformities on the order of 10^{-5} were produced in the course of galaxy formation.) It was followed by a much slower stage of gradual expansion and cooling of the Universe. All particles were relativistic at the beginning of this expansion but a second later only neutrinos (antineutrinos) and electrons (positrons) had velocities comparable to those of photons.

15.2 The cooling down Universe*

The temperature of the universe T and its age t were related at this moment by the formula

$$4\pi^3 G_N T^4 t^2 \approx 1 \qquad (15.1)$$

or $10T^2 t/m_P \approx 1$. We now take into account that $\hbar \approx 6.6 \cdot 10^{-22}$ MeV·c and that $m_P \approx 10^{22}$ MeV and obtain $t \approx T^{-2}$ where t is given in seconds and T in MeV (a simple derivation of this formula can be found e.g. in sec. 27.1 of [14]).

In three minutes the universe cooled to a temperature of $T \approx 10^9$K ≈ 0.1MeV; at that moment it consisted mostly of photons and neutrinos, as well as small amounts (on the order of 10^{-9}) electrons, protons and nuclei of deuterium and helium. This is explained with wonderful clarity and in great detail by Stephen Weinberg in his outstanding book [15]. We are still not certain what produced these tiny baryon and lepton asymmetries which dictated the evolution of the Universe.

15.3   The cosmic microwave radiation (CMB)*

300 000 years later electrons cooled to the temperature of 1 eV and joined nuclei of hydrogen and helium to form atoms of these elements. Photons emitted then survived till our time, having cooled in the process to $2.3 \cdot 10^{-4}$ eV $\approx$ 2.7 K and becoming radio waves. This cosmic microwave background — the CMB — was predicted in the 1940s and discovered in 1965.

15.4   Dark matter*

We know since the 1930s that peripheral stars in the Galaxy move faster than they would have to move if pulled

by only the visible matter of the Galaxy. The explanation was that the Galaxy contains some invisible matter –
dark matter whose nature was unknown but whose amount
is five times greater than that of ordinary matter which
emits light. In the 1960s first candidates for the role of
dark matter were suggested: the so-called mirror particles which interact with mirror photons but do not interact with ordinary photons (see below). Later this role of
the most likely candidates for dark matter shifted to neutral massive superparticles (see below). It should be possible to clarify the nature of dark matter experimentally
in low-background underground laboratories by traces left
by ordinary particles after collisions with particles of dark
matter.

15.5   Dark energy

Dark energy, discovered in 1998, is the term given to the
phenomenon that cannot be caused by any sort of particle.
Namely, observations reveal acccelleration in the rate of
growth of enormous voids between very distant galaxy clus-
ters. The energy of this mysterious accelleration of growth
of empty space (often called antigravitation) is three times
the combined energy of the dark and ordinary matters.

Chapter 16

Quantum Gravidynamics — QGD

16.1 GRT: three canonical effects*

When people mention descriptions of relativistic effects in the theory of gravitation they typically imply Einstein's General Theory of Relativity (GTR, 1916). We know that this theory implied the following three effects which by now became canonical.

1) Deflection of the light rays by the gravitational field which was first observed during a solar eclipse in 1919 and then later in numerous gravitational lenses.

2) Shift of the wavelength of a photon in the gravitational field measured in the 1960s.

3) The secular precession of the perihelion of Mercury known since the 19th century.

In terms of the GTR these effects are explained by the gravitating body distorting space and time around itself; a photon or any other particle or a body move in this curved space-time described by the metric tensor $g_{\mu\nu}$ in accordance with the principle of least action.

16.2 Graviton and QGD*

I need to emphasize, however, that it is for more than half a century now that an alternative, quantum explanation of these purely classical effects was discussed in terms of Quantum Gravidynamics (QGD) which is analogous to Quantum Electrodynamics (QED). In terms of the QGD, the gravitational interaction operates through the exchange of virtual gravitons which are spin-2 massless particles; this is similar to the electromagnetic interaction occurring through the exchange of photons, which are spin-1 particles. All three of the canonical GTR effects are easy to calculate in QGD using perturbation theory. At very low energies characteristic of these effects, perturbation theory works with much higher accuracy in QGD than in QED since in QGD the role of α is played by the very small product of Newton's constant $G_N \approx 6.7 \cdot 10^{-30} \hbar c^5 \text{GeV}^{-2}$ by the squared energy or momentum of the graviton. The smallness of this product explains in a most natural manner why individual gravitons are simply unobservable and the observation of classical gravitational waves is so difficult that they remain undetected by specially built detectors on the Earth. The emisson of gravitational waves explains the observed phenomenon of change of the double pulsar period.

It is very important to emphasize here that Quantum Gravidynamics takes into account not only the so-called graviton ladder diagrams but also such truly loop diagrams in which virtual gravitons interact with one another.

16.3 Nonrenormalizability of QGD

On the other hand Newton's constant G_N has dimension m^{-2}, in contrast to dimensionless α. Consequently, all amplitudes scale as energy squared and the QGD, in contrast to the QED, is a nonrenormalizable theory. As a result of increase in amplitudes, probabilities and cross sections with increasing energies, at high energies perturbation theory becomes unusable. When gravitational interaction becomes strong to the extent that rays of light cease to be straight lines we have to say that the initially flat spacetime is curved by gravitation. We are then left without a means of defining the flat space in the presence of strong gravitational field. Minkowski was able to show in 1908 that in 1905 Einstein introduced flat spacetime into physics. In 1915 Einstein introduced a curved spacetime and predicted three canonical effects of general relativity. It defies imagination how he was able to develop the classical theory of strong gravitational interaction having at his disposal only one tiny perturbative effect (the other two emerged later; in fact, even the first one — Mercury's precession — was not really all that necessary). It goes without saying that the classical theory is not applicable at Planckian distances, time, momentum and energy (see the section "Planck scale" below). Nevertheless, we can use the GRT language and the concept of the metric tensor $g_{\mu\nu}$ in an enormous range of gravitational phenomena.

16.4   Gravitational radius and black holes

The expression for metric tensor $g_{\mu\nu}$ at a distance r from a body of mass m derived by Schwarzschild in 1916 has the form

$$g_{00} = (1 - r_g/r) \tag{16.1}$$

$$g_{rr} = (1 - r_g/r)^{-1}, \tag{16.2}$$

where r_g is the so-called gravitational radius:

$$r_g = 2G_N m \tag{16.3}$$

For the Sun $r_g \approx 3\,\mathrm{km}$ and for the Earth $r_g \approx 1\,\mathrm{cm}$. The Sun's and the Earth's radii are much larger than their gravitational radii. Imagine a compact massive body whose size is less than its gravitational radius. John Wheeler christened such bodies "black holes": regardless of how fast a particle moves, it will never fly beyond the sphere of radius r_g surrounding the black hole.

The biggest black hole in our Galaxy lies at its center.

16.5   The principle of equivalence?

Einstein's famous equivalence principle which states that the gravitational field is indistinguishable from a uniformly accelerated reference frame is valid only so far as the field can be regarded as uniform. The field of any celestial object cannot be uniform as the size of the object is finite. The head and feet of the passenger in the famous falling elevator are subjected to tidal forces because the Earth's gravitational field is nonuniform: it is stronger at the feet level and weaker at the head level. Similar lunar tidal forces cause tides in terrestrial oceans.

The mathematical apparatus of general relativity is definitely correct and is not related to the equivalence principle which holds only in the limiting case, unrealizable in nature, when the gravitational field can be considered perfectly uniform. We should be grateful to this principle as a springing board that launched Einstein to his discovery of the GTR.

16.6   Planck scale

At the turn of the 20th century Max Planck not only introduced the quantum of action but also a combination of the constants G_N, $\hbar$ and c having the dimension of mass:

$$m_P = (\hbar c / G_N)^{1/2}. \qquad (16.4)$$

It became known as the Planck mass m_P.

It is easy to check that $m_P = 1.22089(6) \cdot 10^{19} \text{GeV}/c^2 \approx 2.2 \cdot 10^{-5}$ g.

The Planck energy $E_P = m_P c^2$ and Planck momentum $p_P = m_P c$, equal approximately to $1.2 \cdot 10^{19}$ GeV (GeV/s), as well as the Planck length l_P and the Planck time t_P: $l_P = \hbar/m_P c \approx 1.6 \cdot 10^{-35}$ m and $t_P = \hbar/m_P c^2 \approx 0.5 \cdot 10^{-43}$ s are defined similarly. The gravitational attraction of two particles colliding at an energy above the Planck value may exceed their kinetic energy and result in the generation of a Planck black hole. A theory describing the physics at the Planck scale has not yet been developed. There is no doubt that the construction of nonperturbative quantum gravity would be a tremendously important step forward in the progress of physics.

This concludes our consideration of the gravitational interaction. The next chapter deals with the short-range strong and weak interactions.

Chapter 17

Intranuclear Forces

17.1 Alpha, beta and gamma rays*

On the turn of the 19th century it became clear that some substances — radioactive materials – emit α, β and γ rays. It was then understood that α rays are helium nuclei, β rays are electrons, and γ rays are photons carrying very high energy, higher by many orders of magnitude than the energy of atomic photons.

17.2 Strong interaction*

In 1910 Rutherford and his colleagues discovered atomic nuclei when sending α rays through gold foils. A decade later it was understood that nuclei contain protons; in another ten years, that they contain neutrons. The interaction between protons and neutrons that makes possible the formation of nuclei was called "strong". It is much stronger than the electromagnetic interaction and is described by binding energies measured not in electron-volts but in Megaelectron-volts.

17.3   Isotopic spin

The discovery of the neutron led to the introduction of the concepts of isotopic spin and isotopic invariance of the strong interaction. In the imaginable isotopic space, the proton–neutron doublet corresponds to two projections of the isotopic spin of the nucleon: $+1/2$ and $-1/2$. Mathematicians denote symmetries of the type of the isotopic symmetry by SU(2). The numeral 2 here points to the doublet.

17.4   Weak interaction*

The study of β rays led to the discovery of the weak intranuclear interaction. It was called weak because inside nuclei it is considerably weaker than the electromagnetic interaction and the processes it drives are much slower than electromagnetic processes. The main weak process in the nucleus is the β decay of a neutron n into a proton p. During the 1930s physicists understood that this process involves the emission of an electron e and a neutrino (in fact an antineutrino $\bar{\nu}$) — a particle that for a long time remained hypothetical:

$$n \to p + e + \bar{\nu}. \qquad (17.1)$$

17.5   Neutrino and four-fermion interaction*

The hypothesis of the existence of neutrinos was proposed in 1930 by Pauli. In 1934 Fermi proposed a four-fermion theory in which the β decay is a result of interaction between two weak currents: the nucleon current $\bar{n}p$ and the lepton current $\bar{e}\nu$ with the coupling constant $G_F/(\hbar c)^3 = 1.166\,37(1) \cdot 10^{-5}$ GeV^{-2}. The same interaction should lead to the reaction of formation of a neutron and a positron in the collision of an antineutrino and a proton: $\bar{\nu} + p \rightarrow e^+ + n$. This reaction was first observed in 1956 with a beam of antineutrinos emitted by a nuclear reactor.

17.6 Nuclear fission*

The phenomenon of fission of some heavy nuclei colliding with neutrons into lighter ones was discovered in the late 1930s; this reaction released energies on the order of a hundred MeV. The fission of a nucleus also released free neutrons which could trigger fission in other nuclei; this observation gave rise to the idea of feasibility of a chain fission reaction. This fission reaction was first implemented in a nuclear reactor in 1942 and in an atomic bomb in 1945.

17.7 Nuclear fusion*

It became equally clear by the end of the 1930s that at high temperatures the combination of the strong and weak processes leads to reactions of fusion: energy-releasing merger of nuclei in which protons and electrons transform into nuclei of helium, and then into even heavier nuclei. It was understood that processes responsible for the release of energy in the Sun and the stars are precisely these thermonuclear reactions of nuclear fusion. The first thermonuclear bomb was exploded in 1953.

17.8   From nuclei to particles*

The discovery and consequent study of nuclear interactions has given mankind new powerful sources of energy and allowed physicists to understand why and how stars shine.

The most important consequence of this research was the discovery of elementary particles of matter that we discuss in the next chapters.

Chapter 18

Particles in Cosmic Rays

18.1 Positron*

The first particle discovered in cosmic rays was the positron. Cosmic rays — this flux of particles streaming into the Earth atmosphere from the outer space — were discovered in 1912. In 1932 Anderson discovered that collisions of cosmic rays with matter generate particles with the mass of the electron but with the opposite sign of electric charge. Thus was discovered the first antiparticle that Dirac predicted two years earlier.

18.2 Muon*

In 1938 Anderson and Neddermeyer discovered the muons, $\mu^{\pm}$ particles with a mass of 105 MeV/c^2 produced by cosmic rays high in the atmosphere and reaching the Earth surface even though, as found later, the lifetime of a muon at rest is only 2 microseconds. This therefore was the discovery of the phenomenon predicted by the special theory of relativity: the lifetime of a relativistic particle with energy E and mass m increases by a factor of E/mc^2. This vividly demonstrated that the motion of a particle in flat spacetime lengthens its lifetime.

At the same time, the discovery of the muon was the first step on the way to discovering numerous elementary particles which, to a cursory glance, play no part in either reactors or bombs, nor in stars. Nevertheless, these particles did make it possible to understand how the strong, weak and electromagnetic interactions really work.

18.3   Three pions*

In 1947 came the discovery of strongly interacting charged $\pi^{\pm}$ mesons with the mass of 140 MeV. It was established that they decayed by the channel $\pi \to \mu + \nu$ via the weak interaction and that the muon is a particle that plays no part in the strong interaction — it is a heavy analogue of the electron. The existence of the pion was already predicted in 1935. The exchange of pions between nucleons was supposed to explain nuclear forces. It was also expected that a neutral pion π^0 exists alongside the charged pions. However, it was discovered only in the early 1950s. This particle of mass 135 MeV/c^2 decays to two photons. Three of the pions form an isotopic triplet by analogy to two nucleons forming the isotopic doublet. In 1949 Fermi and Yang proposed a composite model of pions in which they constitute bound states of a nucleon and an antinucleon (not yet discovered at the time).

18.4 Strange particles*

The first so-called "strange" particles — K mesons with a mass of about 500 MeV — were observed in cosmic rays in the same period. They were followed by the first strange baryons given the name hyperons: Λ^0, Σ^+, Ξ^-. These particles were called strange because they were created frequently and fast (through the strong interaction) but then decayed slowly to strongly interacting particles (through the weak interaction). Then arrived the discoveries of the doublets K^+, K^0 with positive strangeness and of the doublet of the corresponding antiparticles K^-, $\bar{K}^0$ with negative strangeness. When created in collisions of non-strange particles, K mesons were created together with anti-K-mesons or with hyperons, so that the total strangeness was conserved.

18.5 Strangeness

Formally, the strangeness was defined by the formula

$$Q = T_3 + B/2 + S/2, \qquad (18.1)$$

where Q is the charge of the particle, T_3 is the projection of its isotopic spin, B is its baryon number (equals $+1$ for baryons, -1 for antibaryons and 0 for mesons) and S is the strangeness. This formula shows that the strangeness of pions and nucleons is zero, the strangeness of K mesons is $+1$, the strangeness of $\bar{K}$ mesons and Λ^0-, $\Sigma^\pm$-hyperons is -1, and that of Ξ-hyperons is -2.

Chapter 19

Particles in Accelerators

19.1 Baryon resonances and antinucleons*

In the early 1950s the study of elementary particles moved to accelerators specially constructed for this purpose.

In 1952 scattering of pions by nucleons revealed the first resonance states — four particles Δ^{++}, Δ^+, Δ^0, Δ^- with masses of about 1230 MeV, which rapidly decayed to a nucleon and a pion through the strong interaction.

In 1955 antiprotons $\bar{p}$ were created in an accelerator. Next year antineutrons $\bar{n}$ were created and observed.

19.2 Sakata model

In 1956 Sakata suggested, in his generalization of the Fermi–Yang model, that the most fundamental among all baryons and mesons known at the time were the isotopic doublet of nucleons and the isotopic singlet Λ^0, while all others were built out of them and of their antiparticles (see e.g. [16]).

19.3   Three sakatons

In 1957–58 another hypothesis was put forward, assuming that all baryons and mesons known at the moment, including also p, n and Λ, are built of three more fundamental particles with quantum numbers of the isotopic doublet and singlet of baryons (and the corresponding antiparticles). These particles — one charged and two neutral — were called the sakatons. In this model all mesons consisted of a sakaton and an antisakaton, and all baryons — of two sakatons and one antisakaton.

Selection rules have been formulated for the strong and weak interaction processes in terms of the sakaton model. Thus a prediction was formulated of the isotopic $SU(2)$ properties of weak currents, both those changing strangeness and those preserving it. A hypothesis was discussed of the existence of a more general symmetry of the strong interaction which was soon denoted by $SU(3)$. This symmetry was assumed broken by the mass of the first two sakatons being smaller than that of the third. It was conjectured that the weak current transforming the first sakaton into the second (and therefore strangeness-preserving) had the weak interaction coupling constant approximately four times greater than the weak current transforming the first sakaton into the third (and therefore not preserving strangeness).

19.4   Octet and singlet of pseudoscalar mesons

The sakaton model predicted the existence of two pseudoscalar mesons with zero isotopic spin which were later called the η and η' mesons. The η meson with a mass of 548 MeV was discovered in 1961. Together with three π mesons and four K mesons they formed the first meson octet within the SU(3) symmetry which was based on the assumption that the masses and strong interactions of the three sakatons were identical. The meson η' was discovered in 1964. Its mass was found to be 958 MeV. This was the first SU(3) singlet.

19.5   Nine vector mesons

In 1961, first experimental data emerged on the nine vector mesons: an isotopic triplet of ρ mesons with $m \approx 770$ MeV, a singlet ω meson with $m \approx 780$ MeV, two doublets K^* and $\bar{K}^*$ with $m \approx 890$ MeV and then the ϕ meson with $m \approx 1020$ MeV.

19.6   Octet of baryons*

Having successfully described the meson octet and a singlet, the sakaton model could not explain the existence of the baryon octet with spin $\frac{1}{2}$: two nucleons ($m \approx 940$ MeV) and six hyperons (one Λ ($m \approx 1115$ MeV), three Σ ($m \approx 1200$ MeV) and two Ξ ($m \approx 1320$ MeV).

19.7 Decuplet of baryons

In addition to four Δ baryons with $m \approx 1230$ MeV with the spin and parity $\frac{3}{2}^+$, a Σ^+ hyperon with $m \approx 1385$ MeV and the same spin and parity was discovered in 1960 and first indication became apparent of the existence of Ξ^- with $m \approx 1535$ MeV. This was sufficient for Gell-Mann to predict in 1962 the existence in the decuplet with $J^P = \frac{3}{2}^+$ of the tenth baryon Ω^- with mass $m \approx 1685$ MeV. This particle was discovered in 1964.

19.8 Conference at CERN 1962

The discovery at Brookhaven National Laboratory of the muon neutrino ν_μ was announced in 1962 at a conference at CERN, and the term 'hadron' was introduced.

19.9 Three quarks*

In 1964 Gell-Mann and Zweig published a hypothesis on the existence of three quarks u, d, s with fractional electric charges ($+2/3$ for u and $-1/3$ for d and s) and the baryon number $1/3$. They were therefore able to explain the existence of the SU(3) octet and decuplet of baryons as bound states of three quarks. For instance, uuu is the $\Delta^+{+}$, ddd is the Δ^-, and sss is the Ω^-. In this model the proton is formed by uud, the neutron by udd and the Λ^0 hyperon by uds.

Chapter 20

Three Discrete Symmetries

The experimental study of K mesons led in 1956 to setting up experiments on probing the bounds on the validity of the three discrete symmetries: C, P, T.

20.1   C, P, T operations*

Consider any process controlled by any interaction.

Let us apply (in our mind's eye) one of the following three operations:

1. Replace all particles involved in the process with their antiparticles. This is known as charge conjugation and is denoted correspondingly by the letter C (from Charge).

2. Reflect the process in a mirror, or reverse the signs of all three spatial axes. (Reversing the sign of one axis and rotation around it by 180° reverses the signs of all three axes.) This operation is known as spatial reflection and denoted by the letter P (from Parity). This operation makes the vectors of position **r** and momentum **p** reverse sign (being odd) while the pseudovectors (also called axial

vectors) of angular momentum **L** and spin **S** do not (being even).

3. Reverse the arrow of time flow. This operation is known as time reversal and denoted by T (from Time).

The electromagnetic interaction is invariant under each of these transformations: they transform any electromagnetic process into another process which can also occur in nature.

It was found out later that the gravitational and strong interactions possess the same property. However, in 1956 it was established that the weak interaction totally violates the P and C, and in 1964 — that it also violates the CP invariance to some extent (see below). If the CPT invariance holds, than this means that the T invariance is broken too. The CPT invariance lies at the foundation of the edifice of Nature that we discuss here. So far experiments detected no events violating the CPT.

20.2 Nonconservation of mirror symmetry P*

In 1956 the fact that the K^+ meson decays to two and three pions made it necessary to scrutinize if weak interactions conserved the P symmetry. Dedicated experiments proposed by Lee and Yang demonstrated early in 1957 that this symmetry breaks in weak processes to the maximum possible degree.

20.3   Nonconservation of charge symmetry C*

It became clear at the same time that weak processes also totally violate the charge symmetry C.

20.4   The hypothesis of conservation of CP symmetry

In 1957 Landau proposed the hypothesis that nature ought to stick to rigorous symmetry with respect to combined CP transformation.

20.5   Conserved vector current

In 1955 Gershtein and Zeldovich advanced a hypothesis of the conserved vector weak current for nucleons and π mesons. According to this hypothesis, the constant of the vector weak interaction which transforms the neutron into the proton is not modified by strong interactions, by analogy to the electric charge of the proton which is not modified by virtual particles. This hypothesis implied that the constant of the β decay of the pion should be greater than the constant for the nucleon by a factor of $\sqrt{2}$. Three years later, this same hypothesis was proposed by Feynman and Gell-Mann. Experiments soon confirmed its validity.

20.6   V-A current

In 1958 Marshak and Sudarshan, Feynman and Gell-Mann, and independently Sakurai conjectured that all weak processes result from the interactions of various components of the universal weak V-A current with its Hermitian-conjugate current.

For example, the electron component has the form $\bar{e}\gamma^\alpha(1+\gamma^5)\nu_e$ where $\bar{e}$ is the wave function of the emerging electron, ν_e is the wave function of the absorbed electron neutrino, $\bar{e}\gamma^\alpha\nu_e$ is the vector part V of the current and $-\bar{e}\gamma^\alpha\gamma^5\nu_e$ is its axial–vector part A. The muon component looks similarly, and in the sakaton model so look the terms $\bar{n}p$ and $\bar{\Lambda}p$. Later sakaton currents were replaced by quark currents $\bar{d}u$ and $\bar{s}u$.

This form of the weak current signifies that in the ultra-relativistic limit in which $E >> mc^2$, only the so-called left-polarized particles whose spin is directed against their momentum take part in the weak interaction. Such particles are said to have left helicity. The helicity of massless particles is an exact quantum number. However, the helicity of a particle with mass is not a Lorentz-invariant concept. Strict helicity is violated by the mass m of the particle.

20.7 Helicity and chirality*

Strong violation of mirror symmetry in the weak interaction resembles the symmetry violation between the left-handed and the right-handed in living nature. I will remind the reader that no spatial translation can make the right and the left hands to identically coincide in space, just as two tetrahedra ABCD and ABDC cannot be maneuvered to coincide in space. They are said to have opposite chirality. The word chirality (from the Greek for hand $\chi\varepsilon\iota\varrho$) was introduced in 1904 by Lord Kelvin. We know that all living matter possesses dissymmetry, i.e. it totally violates the symmetry of the left-handed and right-handed.

20.8   Nonconservation of the CP*

In 1964 the $K_L^0 \to \pi^+\pi^-$ decay was discovered. Previously it was assumed that the long-lived K_L^0 had negative CP-parity while the system $\pi^+\pi^-$ should have positive CP-parity; as a result it was concluded that the CP-parity is broken. Until this result it was assumed that only the short-lived K_S^0 meson can decay into $\pi^+\pi^-$.

20.9   Mirror particles as the first version of dark matter

In order to return to some degree of symmetry between a process and its mirror reflection, a hypothesis was put forward of the existence of so-called mirror particles which possessed neither electromagnetic nor weak, nor of course strong interaction with ordinary particles. However, all these interactions were assumed to work among mirror particles and that the phase that violated the CP symmetry for them had the opposite sign. Obviously, according to this hypothesis, all fermions and bosons should have mirror twins. The only exception was the graviton. The true nature of dark matter has not yet been established in experiments.

Chapter 21

Half a Century Later*

What happened in particle physics in the years that followed the publication of the quark hypothesis?

21.1 Six quarks

I should begin with the rise of the number of quarks from three to six: heavy quarks c, b and t were added to u, d and s. First hadrons containing the c quark were discovered in 1974. First hadrons containing the b-quark were discovered in 1977. Finally, the heaviest quark — the t — was discovered in 1994.

21.2 Six leptons

The number of leptons rose from four to six: the heavy lepton τ was added in 1975 to e, μ, ν_e, ν_μ, and later its neutrino ν_τ.

21.3 Three generations

As a result quarks and leptons formed three generations of fundamental fermions with spin $1/2$:

$$
\begin{array}{ccc}
u & c & t \\
d & s & b \\
\nu_e & \nu_\mu & \nu_\tau \\
e & \mu & \tau
\end{array}
$$

21.4 Electroweak bosons

In 1967 the electroweak theory was created; the theory predicted the existence of electroweak vector bosons: W^+, W^- and Z^0 with the spin of unity. The exchange between charged weak currents carried by W bosons results in the usual β-decay-type weak interactions. The Z boson realizes the new type of the weak interaction between neutral currents. Interactions between neutral weak currents were discovered at CERN in 1973. The special collider was built there in which the W and Z bosons were discovered in 1983.

The last undiscovered element of the electroweak theory was the spin-zero scalar boson. This is called the Higgs boson and is denoted by the letter h. To search for the Higgs, the Large Hadron Collider (LHC) has been built at CERN.

21.5　Gluons

In 1972 quantum chromodynamics was formulated — the theory of strong interactions between quarks, working by exchange of new spin-one particles called gluons and denoted by the letter g. The next two chapters are devoted to a more detailed description of quantum chromodynamics (QCD) and the theory of electroweak interactions.

21.6　All fundamental bosons

Let us list now all fundamental bosons introduced so far here: one spin-2 particle — graviton; five spin-1 particles (the photon, three electroweak bosons and the gluon); and finally, a spin-0 particle — the higgs.

Chapter 22

On Quantum Chromodynamics

22.1   Color and SU(3) symmetry*

The numbers I mentioned at the end of the preceding chapter (six quarks and one gluon) have to be immediately emended. In fact each of the six quarks exists in three different manifestations known as colors. As for the gluon, it exists in eight different color manifestations.

Both these features stem from the existence of the color SU(3) symmetry which is much more fundamental than the approximate symmetry of the three light quarks u, d, s. Color symmetry was discovered as a result of the comparison of a number of very different patterns of strong interactions, established in numerous experiments.

22.2   Color quarks*

We know that there are six quarks, each with baryon number $1/3$ and electric charges $+2/3$ for u, c, t and $-1/3$ for d, s, b. Each quark exists in three color forms: yellow — y, blue — b, red — r. Likewise, antiquarks are "painted" into complementary colors: purple — p, orange — o, green —

g. These "colors" obviously have no connection with the colors of the optical spectrum. It is nevertheless convenient to use them for denoting specific "charges" that characterize the interactions between quarks inside hadrons. Note that hadrons composed of quarks have no color charge. They are said to be colorless or white.

22.3   Colored gluons*

Quarks interact with one another by exchanging eight colored gluons: $8 = 3 \times \bar{3} - 1$. The combination yp+bo+rg is white. All hadrons are white. Consequently, strong interactions between hadrons (including nuclear forces) resemble the forces between electrically neutral atoms. In the same vein, the forces between hadrons are weaker than the true strong color forces between colored gluons and quarks.

22.4   Confinement*

Colour forces possess the remarkable property of confinement. Colored particles cannot break out of white hadrons since they cannot exist in a free states.

Even though nothing of what we have learned so far about hadrons contradicts the confinement hypothesis, we still do not have a consistent quantitative theory of this remarkable phenomenon.

22.5   Masses of nucleons

To put it crudely, the mass of a nucleon equals the sum of kinetic energies of the ultrarelativistic quarks which are locked inside it and by virtue of the confinement have to return into the depth of the nucleon. Their momenta are turned back close to the confinement radius, which resembles the deflection of charged particles by magnetic field.

22.6   Chiral limit*

In the limit when the masses of the light quarks (u, d) can be assumed negligible (and set equal to zero), the mass of nucleons remains practically unchanged. Not so the masses of pions: they change drastically and pions become massless. The states with arbitrary number of zero-energy pions thus become degenerate in energy. As pions are pseudoscalar, this would imply strict conservation of the axial current (CAC — Conserved Axial Current). This limit is known as chiral, because the currents of the left- and right-polarized light quarks would then be separately conserved. In reality pions are certainly not massless but nevertheless are 7 times lighter than nucleons and about 5 times lighter than ρ mesons which also consist of light quarks. This means that the chiral symmetry does hold albeit approximately. This is equivalent to partial conservation of axial currents (PCAC — Partially Conserved Axial Current).

22.7   Pion masses

Experimental data on pion masses allows evaluation of the masses of light quarks $q = u$, d, assuming that

$$m_q/F_\pi = m_\pi^2/m_\rho^2, \qquad (22.1)$$

where $F_\pi = f_\pi/\sqrt{2} \approx 93\text{MeV}$ is the coupling constant characterizing the interaction between pions and the axial current. The validity of this formula is discussed in chapter 19 of Weinberg's book "Quantum field theory" [3]. This evaluation implies that the mass of light quarks is of the order of 10 MeV.

Since the proton is lighter than the neutron we conclude that the u-quark is lighter than the d-quark.

22.8   Masses of other quarks

The data on the masses of strange particles implies that the mass of the s-quark $m_s \approx 100$ MeV. The mass of the c-quark $m_c \approx 1.3$ GeV, that of the b-quark $m_b \approx 5$ GeV, and that of the t-quark $m_t \approx 170$ GeV.

22.9   QCD today

At the moment of writing, QCD is the fundamental tool for the calculation of creation of hadrons at the Large Hadron Collider (see pedagogical discussion of the computation of Feynman tree diagrams (i.e. diagrams without loops) in QED and QCD in [17].

Chapter 23

On the Electroweak Theory

23.1 Intermediate bosons*

Charged weak currents interact by exchanging $W^{\pm}$ bosons. Their mass $m_W \approx 80$ GeV. Neutral weak currents interact by exchanging Z^0 bosons. Their mass $m_Z \approx 91$ GeV. The ratio $m_W/m_Z = \cos\theta_W$ dictates the magnitude of the important parameter of the electroweak interaction θ_W known as the Weinberg angle. The Fermi coupling constant G_F is expressed in terms of the W boson mass as follows:

$$G_F = \pi\alpha/\sqrt{2}m_W^2 \sin^2\theta_W. \qquad (23.1)$$

This formula shows that the smallness of the Fermi constant follows from the large mass of the W boson. Experimentally, $\sin^2\theta_W \approx 0.22$.

23.2 Toy model SU(2) × U(1)

Let us consider a toy model of the electroweak interaction which only has two massless leptons e and ν and four massless vector bosons W^+, W^-, W^0 and B^0. Let

105

W bosons be the components of the isotopic triplet under the group SU(2), and the B boson be the singlet under the groups SU(2) and SU(1). Let left-handed leptons form the isotopic doublet so that $T_3 = 1/2$ for ν_L and $T_3 = -1/2$ for e_L, while right-handed leptons form isosinglets: $T(\nu_R) = T(e_R) = 0$. Assume now that the source of the triplet of W bosons is the isospin T and the source of the B^0 bosons is the quantity $Q - T_3$ equal to 0 for ν_R, to -1 for e_R, and to $-1/2$ for ν_L and e_L. Finally we assume that the the ratio of singlet to triplet "charges" g_1 and g_2 is that of $\sin\theta_W$ to $\cos\theta_W$.

23.3 Photon and Z boson in the model SU(2) × U(1)

Consider now the combination

$$A = B^0 c + W^0 s, \qquad (23.2)$$

$$Z = -B^0 s + W^0 c. \qquad (23.3)$$

corresponding to a photon and a Z boson, respectively. We obtain

$$B^0 = Ac - Zs, \qquad (23.4)$$

$$W^0 = As + Zc, \qquad (23.5)$$

where for brevity we use the notation $s = \sin\theta_W$ and $c = \cos\theta_W$.

The conclusion from the above operations is that e_L interacts with the combination

$$- (B^0 s + W^0 c)/2 =$$
$$- (Ac - Zs)s/2 - (As + Zc)c/2 =$$
$$- Acs - Z(c^2 - s^2)/2, \quad (23.6)$$

and e_R interacts with the combination

$$-B^0 s = -Acs + Zs^2. \quad (23.7)$$

Now we take into account that the vector and axial currents are equal to the sum and difference of the left-handed and right-handed currents. Then the last three formulas imply that the photon interacts only with the vector current of the electron, with coupling constant $-cs$. The Z boson interacts with the electron vector current with a coupling constant $s^2 - \frac{1}{4}$, and with the axial vector current of the electron with a coupling constant $\frac{1}{4}$. $W^\pm$ bosons are related to the charged currents that transform the neutrino into the electron, and the electron into the neutrino with the coupling constant c.

23.4 The first step towards a realistic model

In order to present a description of electroweak interactions of all fundamental fermions we need to complement the doublet (ν_e, e) with two other lepton doublets: (ν_μ, μ) and (ν_τ, τ), plus also quark doublets.

In the case of quarks it is essential that not only the three main doublets with a coupling constant of approximately 1 be considered:

$(u, d), (c, s), (t, b),$

but also six so-to-speak "mixed" doublets providing transitions between the generations:

$(u, s), (c, d),$

with a coupling constant of approximately 0.22;

$(c, b), (t, s),$

with a coupling constant of approximately 0.05;

$(t, d), (u, b),$

with coupling constants smaller than 0.01.

23.5   The second and final step

The semi-toy model described above is good in that it considers in a unified manner the electromagnetic and weak interactions thus offering to describe an immensely broad range of phenomena. At the same time, it is no good in the sense that its W and Z bosons are massless while in nature they are heavier by two orders of magnitude than nucleons. To address this shortcoming, some theorists suggested to use the mechanism of spontaneous breaking of the $SU(2) \times U(1)$ symmetry; the mechanism is known as the Higgs mechanism, by the name of one of the authors who discussed it earlier.

23.6 Doublet of scalar fields

This mechanism is typically explained in terms of the concept of scalar field φ and density potential of this field $V(\varphi) = \lambda^2(|\varphi|^2 - \eta^2)^2$.

Here

$$\varphi = \begin{pmatrix} \varphi^+ \\ \varphi^0 \end{pmatrix}, \qquad (23.8)$$

φ is the isotopic doublet , $|\varphi|^2 = \bar{\varphi}^+\varphi^+ + \bar{\varphi}^0\varphi^0$ is the isotopic scalar, λ is a dimensionless parameter whose value is yet unknown to us.

I will show a little later that

$$\eta = 2^{-3/4}G_F^{-1/2} = 174 \text{ GeV}. \qquad (23.9)$$

The parameter η is the only dimensional parameter of the theory. The masses of all particles are expressed through it. (Note that the mass of the t quark is 172 ± 2 GeV ! The cause of this amazing coincidence of the quantities m_t and η remains unknown.)

23.7 Spontaneous breaking of SU(2) × U(1) symmetry

As a result of breaking of the SU(2) × U(1) symmetry[1] only one of the four scalar fields survives — the higgs field χ,

---

[1]This violates not only the conservation of chirality of particles but also the so-called gauge symmetry which we do not discuss in the book.

against the background of the vacuum condensate η:

$$\varphi = \begin{pmatrix} 0 \\ \chi + \eta \end{pmatrix}. \qquad (23.10)$$

The other three scalar fields join the company of three vector fields W and Z: indeed, in contrast to massless vector fields which only have two components, the massive vector field needs three components. Note that the scalar condensate is electrically neutral; this is why the photon remains massless.

23.8   Condensate and the masses of fundamental particles

As I remarked above, nucleons' masses mostly result from the confinement mechanism and thus belong to the field of quantum chromodynamics. In contrast to this, the scalar condensate η of the electroweak theory gives mass to all fundamental particles: both to leptons and quarks, and also to the vector W and Z bosons, and to the scalar boson H known as the Higgs boson (or simply higgs):

$$m_W = g_2\eta/\sqrt{2} = \sqrt{2\pi\alpha}\,\eta/\sin\theta_W, \qquad (23.11)$$

$$m_Z = m_W/\cos\theta_W, \qquad (23.12)$$

$$m_H = 2\lambda\eta. \qquad (23.13)$$

As for leptons and quarks, the weight of each of these 12 particles equals $f\eta$ where f is a dimensionless constant, different for each particle. For the neutrino it is somewhere near 10^{-13}, for the electron — $3 \cdot 10^{-6}$, and for the t quark it equals unity. Why these constants are as large or as small as they are is a question that can be attacked once the higgs has been discovered.

23.9 The search for higgs*

The search for the Higgs is problem No. 1 of high energy physics for more than thirty years. Attempts to find this particle were made on every one of the most powerful accelerators. Higgs was the primary target of the Superconducting SuperCollider project (SSC) which unfortunately has not been implemented in the USA.

23.10 Large Hadron Collider*

The main hope of establishing if the higgs exists hangs on CERN's Large Hadron Collider (LHC). If the Higgs is discovered, the above simple theory will have been confirmed. If the higgs is not discovered the situation will be even more exciting because it may then be possible to find out on the same collider how the electroweak interaction actually works at energies on the order of 1 TeV and above. In this case the interaction between electroweak bosons should become strong at about 10 TeV.

The circumference of the LHC ring is 27 km, the energy of each of the two proton beams in it is 3.5 TeV. (This is the energy at which it works since March 2010. CERN plans to raise this energy to 7 TeV in 2014.)

23.11   Summer 2011: results of the quest for higgs

Preliminary results of the search for the Higgs in the LHC colliders (in the detectors ATLAS and CMS) and Tevatron (in the detectors CDF and D0) were reported and compared at the European Conference on High Energy Physics in July 2011 in Grenoble[18] and a month later at the 2011 Lepton–Photon conference in Mumbai[21]. So far these bosons have not been detected.

Chapter 24

Supersymmetry

Unlike the previous twenty-three chapters, chapters 24–26 focus on theoretical extrapolations and speculations which so far have no experimental confirmation. The range of these extrapolations extends from the energy available at the Large Hadron Collider to energies on the order of the Planck energy, absolutely inaccessible ever.

The second target of the Large Hadron Collider can be tentatively identified as a search for the hypothetical supersymmetry particles, known as supersymmetric. The concept of supersymmetry (Supersymmetry, SUSY) first emerged in 1971 and is closely tied with that of spin. The number of papers on supersymmetry that theoreticians published in the last 40 years runs into tens of thousands. So far experimenters failed to identify even a single particle predicted in the framework of supersymmetry.

24.1 Spinor generator

The key concept of supersymmetry is that of the spinor generator Q which changes the spin of a particle by $1/2$.

For example, Q acting on the photon transforms it into a hypothetical spin-1/2 particle photino. There is absolutely no doubt that nature has no massless photinos. Hence, nature has no strict supersymmetry, it has to be broken. Likewise, there is no photino with a mass on the order of 1 GeV (typical of light hadrons). Many theoreticians expected the masses of supersymmetric particles to be much lower that 1 TeV.

Setting the aspect of supersymmetry breaking aside for the moment, we begin with an unbroken SUSY. Let us consider the anticommutator of two spinor generators:

$$[Q, \bar{Q}]_+ \equiv Q\bar{Q} + \bar{Q}Q = -2p_\mu\gamma^\mu, \qquad (24.1)$$

where p_μ is the generator of four-dimensional displacement and γ^μ are four Dirac matrices. We see that switching from one particle to another and then returning to the original particle, we find it in a different point in space. Supersymmetry thus generalizes special theory of relativity and makes it more profound.

24.2   Spinor-flavor generators

The geometry and internal symmetries merge if one "clips" to the spinor generator a certain "internal" index $i(1 \le i \le N)$ pointing to the particle of the supermultiplet into which the original particle is transformed. The case of $N = 4$ is a special one. It is the so-called $N = 4$ supersymmetry. The theory has one particle with $J = 1$, four with $J = 1/2$ and six with $J = 0$. On the whole this gives eight bosonic and eight fermionic massless states.

24.3 Summer 2011: Results of the search for light superparticles

The negative results of the searches for superparticles lighter than 1 TeV (and for other manifestations of the "new physics") were reported in summer 2011 in Grenoble and Mumbai [19], [20], [21].

24.4 Prospects

What are the prospects of elementary particle physics once the mechanism of breaking of the electroweak $SU(2) \times U(1)$ symmetry at energies on the order of 1 TeV and distances on the order of 10^{-19} m is understood? If supersymmetry turns out to be unrelated to the problem of electroweak symmetry breaking then it would be entirely possible for breaking to manifest itself only on the Planck scale (see chapter 27) but may even remain forever undetected. In this case the problem that would follow after the electromagnetic and the weak interactions have been unified, should be their unification with the strong interaction.

Chapter 25

Grand Unification

Can we anticipate that the electromagnetic and the weak interactions could be even more tightly unified on still shorter distance scales in the sense that their description would need a single coupling constant while the Weinberg angle would be fixed by some higher symmetry? Can we expect that a strong chromodynamic interaction would join them in the process? In other words, would leptons and quarks form a "unified family" at short distances? What would the spacetime and energy–momentum scales be on which this could be expected?

25.1 Running of three coupling constants

The hope of getting an affirmative answer to these questions is based on the observation that the three constants do not differ from one another so very much at distances on the order of 10^{-15} m and that they manifest a tendency to further convergence. Some theoretical models of this unification were developed in which these constants "run in" to a common value

$$\alpha_{GU} = g_{GU}^2/4\pi \approx 1/40 \qquad (25.1)$$

over distances of order 10^{-32} m. (The index $_{GU}$ here stands for Grand Unification.)

25.2 SU(5) symmetry

The simplest model of grand unification is the SU(5) symmetry. In this model, leptons and quarks of each generation belong to two multiplets: a 5-plet and a 10-plet. The 5-plet in the first generation, for instance, consists of $(3\bar{d}_L, e_L^-, \nu_L)$, and the 10-plet consists of $(3d_L, 3u_L, 3\bar{u}_L, e_L^+)$; the index L stands for the left-handed chirality and the coefficient 3 indicates three colors.

The symmetry group SU(5) has $5^2 - 1 = 24$ vector bosons. Twelve of them are the eight gluons, plus the W^+, W^-, Z bosons and the photon. Twelve others are very massive X and Y bosons with masses of the order of 10^{15} GeV. The charges of six X bosons are $\pm\frac{4}{3}$; in the 5-plet they are responsible for transitions between the three $\bar{d}$ antiquarks and the electron and vice versa. The charges of six Y bosons are $\pm\frac{1}{3}$; in the 5-plet they are responsible for transitions between the three $\bar{d}$-antiquarks and the neutrino and vice versa.

25.3 Proton and neutron decays

Consider now the transitions in the 10-plet involving X and Y bosons. Here X bosons can also transform u-quarks into the $\bar{u}$ antiquark and vice versa while Y bosons can also transform u quarks into the $\bar{d}$ antiquark and vice versa.

This implies, however, that a pair uu can transform through X into $e^+\bar{d}$ and a pair ud can transform through Y into $\bar{\nu}\bar{d}$. We have thus obtained the transformations $p \to e^+$ and $n \to \bar{\nu}$.

It goes without saying that the law of energy and momentum conservation does not allow a lone nucleon to transform into into a lone lepton. Therefore what we mean here are decays of the type $p \to e^+\pi^0$, $p \to e^+\pi^+\pi^-$, $n \to \bar{\nu}\pi^0$, $n \to e^+\pi^-$ etc. Ordinary stable atomic nuclei would then also be unstable, with the expected lifetimes on the order of 10^{30} years.

The search for such decays in huge multikilotonne detectors failed to find them at the level of 10^{32} years, showing therefore that the simplest SU(5) model is not realized in nature.

25.4   Other symmetries

The idea of grand unification was examined with the use of other, more complex symmetry groups: from SO(10) to the highest of the exceptional groups E_8. Some of these models have stimulated a search for new exotic processes, such as neutron–antineutron oscillations in vacuum. So far, however, this search yielded no new experimental discoveries. Consequently, I forgo a discussion of other symmetries in this brief text.

Chapter 26

In the Vicinity of the Planck Mass

The characteristic energy of grand unification ($\approx 10^{16}$ GeV) is only three orders of magnitude smaller than the Planck mass around which the gravitational interaction becomes stronger than the electroweak and strong interactions. Here therefore arise conditions for SuperGrand Unification of all particles and interactions in the framework of supergravity.

The multiplet $N = 8$ of supergravity contains: 1 spin-2 graviton, 8 spin-3/2 gravitinos, 28 ($= 8 \cdot 7/1 \cdot 2$) spin-1 bosons, 56 ($= 8 \cdot 7 \cdot 6/1 \cdot 2 \cdot 3$) spin-1/2 fermions and 70 ($= 8 \cdot 7 \cdot 6 \cdot 5/1 \cdot 2 \cdot 3 \cdot 4$) spin-0 bosons; in all, 128 boson and 128 fermion massless states. (Note that spin-zero particle has one state, while spin-non-zero massless particle has two states.)

26.1 Superstrings

Since the early 1980s, most theoreticians working in Planck-scale physics have focused on superstring theory. Superstrings are hypothetical one-dimensional objects of size l on the order of the Planck length $l_P = 1/m_P \approx 10^{-35}$ m, with characteristic tension (energy per unit length) on the order of m_P^2 where m_P is the Planck mass. The ground state of a massless superstring corresponds to massless particles (or practically massless on the m_P scale). The excited states form an infinite spectrum of energy levels of particles with characteristic step of m_P.

The prefix "super" indicates that the spectrum of particles described by the superstring possesses supersymmetry: the numbers of fermion and boson excitations are identical, and their masses are degenerate. Unclosed, open superstrings ("rods") correspond to spin-1 and spin-1/2 particles. Closed superstrings ("rings") correspond to spin-2, 3/2, 1, 1/2 and spin-0 particles.

26.2   Ten spatial dimensions

To construct a superstring theory, it proved necessary to add another six (seven since the mid-1990s) extra spatial dimensions at every point of our space but in contrast to $\mathbf{r}$ and t, they are not linear ones but compacted to a radius of curvature of the order of Planck length. The extra spatial dimension was first introduced by Kaluza in 1919 in order to unify Maxwell's classical electrodynamics and Einstein's general relativity. The attempt proved unsuccessful, just as Weyl's attempt at about the same time to introduce classical gauge (scale) symmetry. By the mid-1920s, the additional spatial coordinate was used by Klein to construct a relativistically invariant quantum equation and by Fock to establish that this equation for electrically charged particles possessed quantum phase invariance which Weyl rechristened several years later into gauge invariance.

26.3 M-theory

In the mid-1990s, it became clear that various models of superstrings could be synthesized within a unified theoretical framework known as the M-theory. It includes, in addition to one-dimensional strings, also multi-dimensional membranes known as branes. Work on the M-theory continues.

26.4 Anti-de Sitter

The name of the mathematician and astronomer de Sitter (1872–1934) is closely connected in modern cosmological literature with the concept of empty non-flat (curved) spacetime. If curvature is positive, this spacetime is called de Sitter spacetime (dS); it corresponds to a positive cosmological constant Λ. If both curvature and Λ are negative, the spacetime is said to be anti-de Sitter (AdS). People usually refer to it as five-dimensional superspace AdS_5 into which our four-dimensional world is embedded. Such concepts are introduced in order some day to build the quantum theory of all forces — the Theory of Everything (TOE). Unfortunately it is not clear to me why this theory should be based on a purely classical concept of a curved spacetime.

Chapter 27

Concluding Remarks*

27.1 "The sun rises, the sun sets."[1]

Recent surveys among the populations of Russia and the United States revealed that in both countries many people cling to the opinion that not only the Moon but equally the Sun rotates around the Earth. The fact that the Sun rises and sets (revolves around us) is not an error but rather a superficial truth which in many ways holds and is sufficient for many. The notion that the Earth is flat is largely sufficient for a person who never went on a long journey. People on an intercontinental flight have a chance to witness that the Earth is spherical but they can continue thinking that the Sun is the Earth's satellite. However, more profound truths exist: the Earth is a sphere, the Moon is a smaller sphere revolving around the Earth, and together they revolve around the Sun.

Superficial truths are adequate for everyday existence. They are absolutely inadmissible, however, in teaching

---

[1] The first line of an old Russian folk song: "The sun rises, the sun sets/but it's always dark in my prison cell etc."

modern astronomy and physics. When a professor tries to persuade his students that the mass of the proton in the Large Hadron Collider increases by a factor of thousands, he is not merely preaching a superficial truth about everyday life — he inculcates into the minds of his students false notions about the theory of relativity, basing them on what passes for the "famous Einstein's formula", $E = mc^2$ (see [25], [26]).

27.2 On teaching of physics

Current quantum and relativistic technologies demonstrate unequivocally that nature has a solid quantum and relativistic foundation. This is the main lesson we learned in the 20th century. More and more people realize that we live in the times of quantum and relativistic civilization. Consequently, simple unclouded picture of the world not contradicting today's physics should be accessible to the maximum number of people from the earliest age even if they will never grow to be physicists. There is no other way to avoiding global catastrophe.

27.3 On the tragic fate of the SSC

It became clear in the last decades of the 19th century that understanding the Higgs sector of the Standard Model is the major unresolved problem of high energy physics. In 1990, the regularly published "Review of particle properties" printed on p. III.12 information on plans of launching three proton–proton colliders: 1) Accelerating-and-Storage Complex UNK (3 TeV, 1995?), 2) Large Hadron Collider LHC (8 TeV, 1996?), 3) Superconducting SuperCollider SSC (20 TeV, 1999) whose task it would be to find a solution to this problem. Launching dates of the first two carried a question mark. No question mark followed the third — Superconducting Supercollider SSC which was to have the highest energy of all — 20 TeV in each beam. The construction of the tunnel for this collider — a ring with circumference of 87 km — was going forward full tilt at the town of Waxahachie near Dallas. Had this project been implemented as planned, by now the Standard Model of elementary particles would probably have ceased to be a model and matured to a complete theory. An even more exciting scenario was on the cards if experiments pointed to a "new physics" stretching beyond the bounds of the Standard Model. Alas, in 1993 the US Congress during the discussion of the budget for 1994, the emerging flagship of high energy physics was scuttled. This delivered a devastating blow to the progress of elementary particle physics (see Clinton's letter [22] which was never given the status of a White House press release). The discussion of the budget was described In Clinton's autobiography [23] but the SSC was not mentioned. In 2011, however, he devoted the final part of his lecture in Davos to SSC(LHC) (see [24]).

A year or two before the termination of the SSC, the construction work in Russia on Accelerating-and-Storage

Complex UNK was discontinued; the complex was to collide proton beams with energy of 3 TeV per beam. Among all TeV-machines, only the TeVatron in the USA worked successfully for many years (1985–2011). It collided beams of protons and antiprotons with energies of 1 TeV.

27.4 LHC and prospects

As I mentioned in the chapter on the electroweak model, the search for the higgses is now conducted at the Large Hadron Collider LHC. So far higgses have not been discovered. There is a chance that the LHC will find and identify them in the near future. This would certainly be a triumph of the electroweak model. In a certain sense, however, it will be much more interesting if LHC data proves that there are no higgses with masses below 1 TeV. This would mean that at energies considerably higher than 1 TeV, the interaction between W (and Z) bosons is truly strong and it is impossible to describe their behavior in terms of the perturbation theory. To study such interactions experimentally, an SSC-scale collider or one still more powerful would be required. The current electroweak model based on perturbation theory uses about two dozens of parameters whose numerical values we are unable explain (i.e. cannot calculate). Perhaps this would become possible by going beyond the confines of perturbation theory.

27.5   On the gist of science and on truth in science

It seems to me that the majority of people fail to grasp the core of the modern science not because it is unfathomable. Of course, it is beyond anyone to understand a subject to its minutest details. Nevertheless, it is possible and even necessary to grasp correct ideas about nature. It is certainly possible to give a pupil an explanation of how the world around is organized and functions once the principal ideas of modern physics have been identified and selected.

Unfortunately, professors fail to agree on which ideas in physics are the most important. I am astonished by the inability of many professional physicists to acknowledge that the mass of special theory of relativity is independent of velocity, and among those for whom the independence of mass from velocity is beyond doubt, their tolerance to having their children and grandchildren taught as before that mass depends on velocity.

I am baffled even more by the almost universal attitude towards quantum mechanics as a prism which shows us an illusory, rippled, unsteady world. In fact it is quantum mechanics that is responsible for the sturdy and steady world surrounding us.

I regard as biased the attitude of many physicists to problems of gravitation and gravitons. It is very fashionable these days (see e.g. [27],[28]) to seek explanations of everything — from the properties of elementary particles to phase transitions — in terms of black holes and superstrings. I do not exclude that some day such a Theory of Everything (TOE) may be constructed but as we see it now, too much of it remains unfinished [29] and we still have no experimental data confirming the validity of the TOE. Consequently, school children should not be introduced to physics by studying this half-baked theory.

What schoolchildren do need as a starting point is the introduction to rigorously established basic constants that nature uses — the velocity c and the quantum $\hbar$; it is not only possible but also very necessary. Indeed, the laws written with these constants do explain the main natural phenomena.

It is very important to understand that the truth exists regardless of whether we will ever be able to achieve complete understanding. And that the target for science is to strive for moving closer and closer to this truth.

Postscript I

Soon after the text of the "ABC of Physics" in Russian was completed and mailed to the translator, a preprint by collaboration OPERA [30] appeared on September 23, 2011 in which some one hundred and fifty authors representing several dozen institutions announced that the velocity of the muon neutrinos v_μ measured in the CERN–Gran Sasso beam was higher than the speed of light in vacuum c. According to [30], $(v - c)/c = 2.48(0.28)[0.30]10^{-5}$ where the number in parentheses () indicates statistical uncertainty and that in brackets [] is the systematic error. According to [30], neutrinos cover the distance of 730 km between CERN and Gran Sasso faster by $60.7(6.9)[7.4]$ nanoseconds than if their velocity equaled c. Light seems to lag behind the neutrino beam by approximately 20 m.

On my first cursory reading of the preprint I did not understand how the time of birth of a neutrino in the pion decay in the kilometer-long CERN tunnel could be established with this accuracy using the GPS satellite system. After more attentive reading I realized that what the system measured was essentially the time between the moment of detection of the proton bunch in the CERN accelerator ring before the beam was spilled onto the target in which protons produced pions, and the moment of detection of neutrinos at Gran Sasso. This was a legitimate procedure as all particles move at velocities close to c. Unfortunately the data of the muonic detector at CERN are not discussed.

The article proved to be hardly readable: it is overloaded with dozens of acronyms which dwarf and sink the scant explanations. It is essential to have a publication with detailed clarifications and a careful analysis of the possible systematic errors.

I will only note that all the systems and high-precision instruments used by the authors are based on the theory of relativity and quantum theory. A paper aspiring to testing relativity theory calls for a different manner of presentation.

Postscript II

To conclude this brief guide over physics, I would like to answer those of my colleagues who sent in their critical remarks about the manuscript; I expressed my deep gratitude to them in the Preface. They all agreed that such a guide is very welcome; however, quite a few of them are of the opinion that the text I've written is far too brief, that it needs considerable expansion, that it should have paid tribute to a greater number of outstanding physicists who contributed to the evolution of physics, and that the bibliography should have been expanded too. On the other hand, most of them believe that the text should have been aimed at a more popular level and that it should have included illustrations. Regretfully, I cannot meet these largely fair and valid wishes in the time that I have.

Bibliography

[1] S. Weinberg, Gravitation and Cosmology: Principles and Applications of the General Theory of Relativity. John Wiley 1972

[2] S. Weinberg, The Quantum Theory of Fields, Volume I. Foundations. Cambridge University Press 2000

[3] S. Weinberg, The Quantum Theory of Fields, Volume II. Modern Applications. Cambridge University Press 2001

[4] Ya. B. Zel'dovich, Higher mathematics for beginners and its applications in physics. Moscow. Mir. 5th edn. 1973

[5] R. P. Feynman, R. B. Leighton, M. Sands, FEYNMAN LECTURES ON PHYSICS, Volume 2, Chapter 19. Addison Wesley, 1964

[6] K. Nakamura et al. (Particle Data Group), J. Phys. G 37, 075021 (2010), Periodic Table
http://pdg.lbl.gov/2010/reviews/
rpp2010-rev-periodic-table.pdf

[7] J Loschmidt
http://www.chemteam.info/Chem-History/
Loschmidt-1865.html
J Perrin
http://nobelprize.org/nobel_prizes/physics/
laureates/1926/press.html

[8] A.I. Akhiezer, V.B. Berestetskii, Quantum Electrodynamics. John Wiley, 1965

[9] N.N. Bogolyubov, D.V. Shirkov, Introduction to theory of quantized fields. John Wiley, 1980

[10] K. Nakamura et al. (Particle Data Group), J. Phys. G 37, 075021 (2010), Table 1.1
http://pdg.lbl.gov/2010/reviews/
rpp2010-rev-phys-constants.pdf

[11] C. W. Allen, Astrophysical Quantities. Springer, 2001 (Units: § 12, Mass, luminosity, radius and density of stars: §, Galaxy: § 135)

[12] Richard Powell, An Atlas of The Universe
http://www.atlasoftheuniverse.com/index.html

[13] L.B. Okun, K.G. Selivanov, V.L. Telegdi, Gravitation, photons, clocks. UFN **169** (1999) 1141-1147
http://www.itep.ru/theor/persons/lab180/okun/
list25/8selivanovtelegdi99.pdf
L. B Okun, K. G. Selivanov, V. L. Telegdi, On interpretation of the redshift in a static gravitational field. American Journal of Physics **68** (2000) 115-119
http:
//www.itep.ru/theor/persons/lab180/okun/em_13.pdf
L.B. Okun, Photons and static gravity. Modern Physics Letters A **15** (2000) 1941-1947
http:
//www.itep.ru/theor/persons/lab180/okun/em_14.pdf

[14] L.B. Okun, Leptons and quarks. Elsevier Science, 1985 ch. 27 Particles of the Universe

[15] S. Weinberg, The first three minutes. Basic Books, Inc., Publishers 1977

[16] Proceedings of International Simposium $pn\Lambda50$ The Jubilee of the Sakata Model. November 25–26, 2006, Nagoya, Japan. Editors M. Harada, Y. Ohnuki, S. Sawada, K. Yamawaki. Progress of Theoretical Physics, Supplement No 167, 2007, which contains a contribution by L. B. Okun, The Impact of the Sakata Model (arXiv: hep-ph/0611298).

[17] Michael E. Peskin. Simplifying Multi-Jet QCD Computation. arXiv:1101.2414

[18] Results on higgs at Grenoble
LHC searches for higgs:
http:
//indico.in2p3.fr/materialDisplay.py?contribId=
985&sessionId=16&materialId=slides&confId=5116
Tevatron searches for higgs:
http:
//indico.in2p3.fr/materialDisplay.py?contribId=
984&sessionId=16&materialId=slides&confId=5116

[19] New physics at Grenoble
new physics at Tevatron
http:
//indico.in2p3.fr/materialDisplay.py?contribId=
953&sessionId=16&materialId=slides&confId=5116
highlights and searches at CMS
http:
//indico.in2p3.fr/materialDisplay.py?contribId=
954&sessionId=16&materialId=slides&confId=5116
highlights and searches at ATLAS
http:
//indico.in2p3.fr/materialDisplay.py?contribId=

      955&sessionId=16&materialId=slides&confId=5116

[20] All talks at the Grenoble Conference
     http://indico.in2p3.fr/contributionListDisplay.
     py?trackShowNoValue=1&OK=1&sortBy=
     number&sessionShowNoValue=1&selTracks=
     0&selTracks=1&selTracks=2&selTracks=3&selTracks=
     4&selTracks=5&selTracks=6&selTracks=7&selTracks=
     8&selTracks=9&selTracks=10&sc=21&selTypes=
     4&selSessions=6&selSessions=8&selSessions=
     9&selSessions=10&selSessions=11&selSessions=
     13&selSessions=14&selSessions=3&selSessions=
     1&selSessions=12&selSessions=15&selSessions=
     16&order=down&confId=5116#contribs

[21] M. Peskin, Summary talk at Lepton–Photon 2011
     http://www.ino.tifr.res.in/MaKaC/getFile.py/
     access?contribId=176&sessionId=23&resId=
     0&materialId=slides&confId=79

[22] President William J. Clinton on SSC (June 1993)
     http://www.presidency.ucsb.edu/ws/index.php?pid=
     46703#axzz1UqKNr5Bs

[23] Bill Clinton, My life. Alfred Knopf, N Y 2004

[24] William J. Clinton speaks at Davos 2011 about LHC. The
     last three minutes (54:30 - 57:00)
     http://www.youtube.com/watch?v=p2dT7xVS6-s

[25] Lev B Okun, Energy and Mass in Relativity Theory. World
     Scientific. 2009.

[26] L. B. Okun, The "relativistic" mug. arXiv: 1010.5400
     Published in "Gribov-80 Memorial Volume: Quantum
     Chronodynamics and Beyond" Yu L Doksitzer, P Levai, J
     Nyri Editors. World Scientific. 2011, pp 439-448.

[27] Brian Green, The Hidden Reality. Alfred A Knopf, N Y 2011
Lecture at Boston Museum of Science March 2 2011
http://www.youtube.com/watch?v=fJqpNudIss4&NR=1

[28] Brian Green, The Fabric of the Cosmos
http://www.youtube.com/watch?v=4SYsDst5Sz0
http://www.youtube.com/watch?v=5u_rYMdBfDo

[29] S. Weinberg, Particle Physics, from Rutherford to the LHC,
Physics Today August 2011 29–33

[30] T. Adams et al, Measurement of the neutrino velocity with OPERA detector in the CNGS beam. arXiv:1109.4897

Some Acronyms

CAC	—	Conserved Axial Current
CERN	—	Organisation Européen de Recherche Nucléair
GTR	—	General Theory of Relativity
JINR	—	Joint Institute for Nuclear Research
LHC	—	Large Hadron Collider
PCAC	—	Partially Conserved Axial Current
QCD	—	Quantum Chromodynamics
QED	—	Quantum Electrodynamics
QGD	—	Quantum Gravidynamics
SSC	—	Superconducting Super Collider
UNK	—	Accelerating-and-Storage Complex

Index